Engineering

GCSE

Engineering

GCSE

Neil Godfrey

Steve Wallis

Published in 2004 by:
Nelson Thornes Ltd
Delta Place
27 Bath Road
CHELTENHAM
GL53 7TH
United Kingdom

05 06 07 08 / 10 9 8 7 6 5 4 3 2

A catalogue record for this book is available from the British Library.

ISBN 0 7487 8551 5

Page make-up by GreenGate Publishing Services, Tonbridge, Kent

Printed in Great Britain by Scotprint

For Hannah

Acknowledgements

The authors would like to thank the following people for their kind help in producing this book:

Our good friends and colleagues Mark Elliott, Michael Casey, Anthony King, Ian Jones, James Shanks, John Barron, Keith Bradley, Adam Turner, Mark Brodie, Phil Shaw and Michael Bretherick for their kind help, support and technical knowledge.

Peter Higgs at EdExcel and Paul Turnbull at SEMTA for their help in reviewing the original manuscript.

Special thanks to Chris Purnell at Nissan UK, Roy Kelly at North East Press Ltd and John Richmond and Graham Robertson at RPC Cresstale.

Finally, special thanks also to Carolyn Lee at Nelson Thornes for her encouragement and support and Katie Chester at GreenGate Publishing for her hard work and patience.

ix

- Rodney Sizemore, p. 89 left
- Rolls-Royce plc, p. 84
- Sony, p. 229
- Tat Ming, p. 112 bottom
- University of Hawaii Institute for Astronomy, p. 199
- Versa Iron & Machine, Inc. St. Paul, Minnesota, www.versaladlecups.com, p. 185 bottom left
- Wilson Process Systems, p. 246

Microsoft screenshots reprinted with permission from Microsoft Corporation. Microsoft and its products are registered trademarks or trademarks of Microsoft Corporation in the United States and/or other countries.

Every effort has been made to contact copyright holders and we apologise if any have been overlooked.

Introduction

Take a look at the products pictured below. Can you guess what they all have in common?

All the products featured, even though vastly different from one another, need to be engineered and manufactured.

Without engineering, you would not be able to watch the latest DVDs, listen to your favourite CDs, play sport in modern training shoes, be able to travel by modern transport or even cook your lunch.

In fact just about every activity that you undertake each day can be related to engineering and manufacturing.

How this book is organised

This book focuses on three important aspects of engineering that you have to study according to the specification. They are:

- Design and Graphical Communication
- Engineered Products
- Application of Technology.

Jargon Dragons

Jargon Dragons in each unit are used to define some important terms, words and phrases which may be unfamiliar to you but are important to understand in engineering.

Activities

Short exercises are used throughout this book to help you understand many of the important themes and topics. The majority of the exercises are designed for you to carry out on your own; however you may require assistance from your teacher/lecturer at various key points.

Product investigation

Product investigations are included at the end of Unit 3. They are designed to help you research key points related to new technology.

Case studies

The case studies outline how real engineering and manufacturing companies produce products in the context of the themes discussed in this book. The case studies are followed by some questions designed to reinforce the key points made.

Find it out

This feature includes research topics and key points for you to investigate and therefore further your understanding of the content.

Think it Through

The Think it Through feature is designed for you to develop your research skills in Units 1 and 2 and to help you revise key points within Unit 3.

Steve Wallis and Neil Godfrey
2004

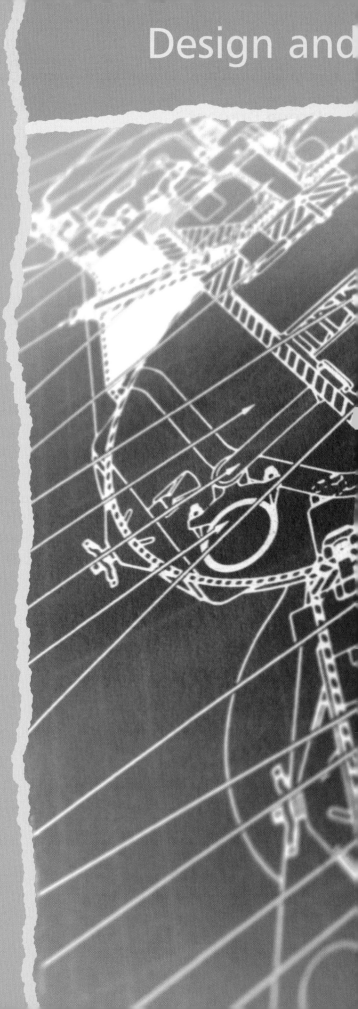

What's in this unit?

To complete this unit you will need to design products from a given design brief. You will learn how to analyse a design brief and to recognise the key features. You will develop a design specification from which you will generate a number of ideas and solutions.

You will learn how to use a range of drawing, design and graphical communications techniques and be able to identify appropriate drawing techniques in order to communicate your ideas to a range of clients.

In your designs, you will need to consider scientific principles, cost and other features to make sure that you have produced the best design to meet the client's needs.

In this unit you will also learn how to evaluate your ideas against the design brief throughout the development of your product's design, to ensure that the final design is appropriate.

Graphical Communication

1

In this unit you will learn about:

Design briefs

A **design brief** is a basic description of what the customer or client wants. A **client** can be a customer or a friend or anyone who is asking the designer to produce a design.

In business, commerce or industry a company that requires a new design will approach a **designer**. The designer may be within the company, a different department within the business or it may be a separate design company.

The design brief is an outline of what the client is looking for. It gives some background to a problem and why the problem needs to be solved. It may give a list of the benefits of designing and producing a new product. It could also outline some of the objectives that the design should achieve.

This information needs to be broken down into important areas known as **key features** and these are as follows:

THE JARGON DRAGON

key features – the important individual parts of a product that need to be considered

- ✔ function
- ✔ quality standards
- ✔ styling aesthetics
- ✔ performance
- ✔ intended markets
- ✔ size
- ✔ maintenance
- ✔ production methods
- ✔ cost
- ✔ regulations
- ✔ scale of production.

This section gives a brief description of each of these key features.

Function
The designer needs to understand what the product is to be used for, where it will be used and what it will need to do.

Quality standards
Most products have to be designed to meet the requirements of **quality standards**. These can include physical characteristics of the product or safety features. The designer

will therefore need to consider all of the relevant standards that relate to the new product.

Styling aesthetics

This term relates to the appearance of the product. Will the product be visually pleasing? Styling can include colour, shape or surface finish. It may include lights or logos on the product that give it an appearance that a customer finds attractive. Products can be designed to have a particular **style**, such as 'futuristic', 'rustic' or 'sleek', which will blend in with other products of the same style.

Performance

The new product will need to perform a task. The designer will need to consider how well the product will do the task, and how long the product will last.

Intended market

The product will be aimed at a particular group of customers. The designer will need to consider this group of customers (**the market**) and look at the types of things that they desire so that the new product will appeal to them.

Size

The designer will need to consider the size of the product. It may be that being small is important, such as with a mobile phone. It may be that being big is desirable to give the feeling of quality and value, for example for a car or a house.

Maintenance

Products do not last forever. Products need to be serviced, fixed or have their components changed from time to time. This is known as **maintenance**. The designer will need to consider how the product will be maintained.

Production methods

How the product will be produced will determine some aspects of the design. The designer will need to understand a range of production methods and how they affect designs.

Often a company will already have a particular type of machine which will need to be used to make the product. Sometimes the product can be designed then the manufacturing process found to best suit the design.

Cost

Modern market forces mean that there is strong competition between companies who produce similar products. In order to sell their products and make a profit they need to make them in the most cost-effective way possible. Some products demand low prices such as nuts and bolts. For some products, such as mobile phones, trainers, CDs and TVs, being expensive gives the customer a feeling of quality and value.

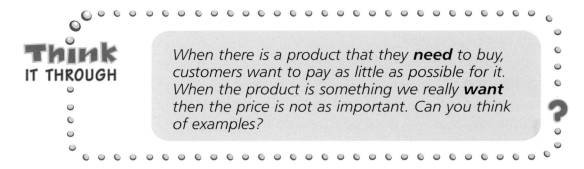

Think
IT THROUGH

*When there is a product that they **need** to buy, customers want to pay as little as possible for it. When the product is something we really **want** then the price is not as important. Can you think of examples?*

Regulations

Most products are designed to meet guidelines. These could be legal obligations of the design and may include restrictions on the use of materials, or safety regulations such as age restrictions.

Scale of production

The design of the product will be influenced by the scale of production. The product may be a one-off, or it may be mass produced.

Developing the design specification and solutions

The designer analyses the design brief (which involves reading it and fully understanding it) so that he or she is able to give a clear explanation of what is required. They will then produce a **design specification**.

There is no standard format for specifications but there are guidelines which should be followed. The **British Standards Institution** (BSI) has produced the document BS 7373:1998; *A*

Guide to the Preparation of Specifications (ISBN 0 580 19005 6), which can be purchased from BSI.

The **key features**, listed on page 4, should be considered when producing a design specification. However, there are many other important parts to a design specification, known as **elements**. You will not need to look at all of these. These are the areas that could be considered when producing a design specification:

- aesthetics
- company constraints
- competitors
- cost of product
- customers
- disposal
- documentation
- environment
- ergonomics
- installation
- legal issues
- life (service)
- life (shelf storage)
- life span of product
- maintenance
- manufacturing facility
- market constraints
- materials
- packaging
- patents
- performance
- politics
- processes
- quality and reliability
- quantity
- safety
- shipping
- size
- standards specifications
- testing
- time scales
- weight.

THE JARGON DRAGON

British Standards – a set of instructions and guidelines used to carry out engineering activities to the correct level of performance and quality

design specification – a list of conditions that must be met. It takes into account the original design brief but also takes into account the decisions made about the key features

So we can see that, as well as the key features, there may be other areas to consider.

A GCSE product specification could be laid out as follows:

- title
- function
- quality standards
- styling aesthetics
- performance
- intended markets
- size
- maintenance
- production methods
- cost
- regulations
- scale of production.

Produce a design specification for the following design brief.

Design brief
A local mountain bike group would like a bicycle rack that will hold two mountain bikes at one time; these are to be carried to various events around the country.

The bike rack should fit a range of cars and it should be easy to attach the bikes to the rack. The design should be robust and weather proof and have a good appearance on the car. The product should be secure from theft and protect the mountain bikes from theft.

Solution

Technical design specification

Title:	Bike rack
Function:	A device to attach mountain bikes to a car for transportation.
Quality standards:	The bike rack must be designed to meet all relevant standards.
Styling aesthetics:	A range of colours including metallic finishes. No square edges. Smooth and curved shape. Maximum size 0.7 m wide.
Performance:	Hold up to two mountain bikes at the same time. Fit to a range of cars. Easily attached and removed. Weather proof. Stable at 70 mile/h. Stable over rough ground.
Intended markets:	The design should be for mountain bike enthusiasts of any age group who would like a stylish, modern rack for their bikes.
Size:	Maximum 750 mm wide × 650 mm high.
Ergonomics:	Lightweight – 10 kg maximum weight. Few moving parts. No sharp edges. No protruding edges.
Maintenance:	The product should be easy to wash. Maintenance should be low cost and easy.
Production methods:	The product should be made from tubular and flat steel which is welded together.
Cost:	Maximum cost £25.00.
Regulations:	The design should met all UK regulations that relate to this type of product.
Scale of production:	The design will be manufactured by small-scale production in batches of 10 products.

Using design techniques

Once the design specification is complete, the designer has clear guidelines to work from. The designer can produce a number of concept drawings that meet the design specification.

Research and analysis of information and data

A new product will have an inter-relationship with other products. It could be attached to something, fit into something or sit on something. It may be part of a complex piece of machinery or it may need to be in a particular environment. It is therefore necessary to consider these external aspects when designing.

Example

A mobile phone holder is to be designed. The holder is to accept a wide range of modern mobile phones. The handset is to fit into the hole with the screen visible.

Data has been collected from a range of handsets.

	Length	Height A	Width	Depth
Handset 1	120	50	35	19
Handset 2	110	40	35	17
Handset 3	65	30	32	22
Handset 4	80	35	37	23
Handset 5	80	34	32	17
Handset 6	75	33	34	19
Handset 7	100	42	40	17
Handset 8	85	40	32	16

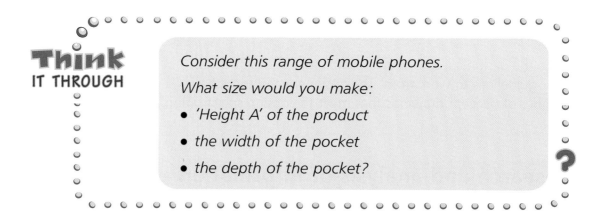

Think
IT THROUGH

Consider this range of mobile phones.

What size would you make:

- 'Height A' of the product
- the width of the pocket
- the depth of the pocket?

Consideration of scientific principles

When products are designed on paper or with **computer-aided drawing (CAD)** systems, the designer has to take into account basic scientific rules so that the product is able to work in the 'real world'.

Some areas that need to be considered are:

- components need to be supported
- levers and gears can magnify forces
- different types of force
- friction
- structures.

Components need to be supported
In the drawing below, a stool is being designed.

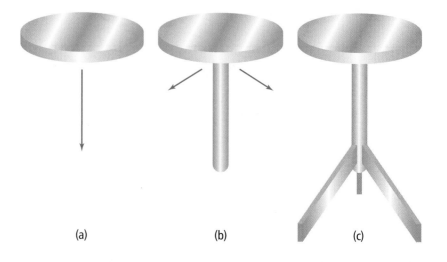

Support development of a simple stool

(a)   (b)   (c)

(a) The seat cannot float in the air. There is a constant force trying to push the seat towards the ground. This is the force of gravity. The seat must be supported by a force that is equal to the force of gravity. The material must be strong enough to resist the force.

(b) A vertical shaft has been added. This must capable of supporting the seat. The designer needs to take into account variation in the forces that will be applied to the seat, such as different-sized people sitting on the stool, and the sort of environment that the stool will be in.

The stool is unstable and will fall over if the centre of gravity moves outside the base. This diagram shows that when the centre of gravity moves past the base of the product it will fall over.

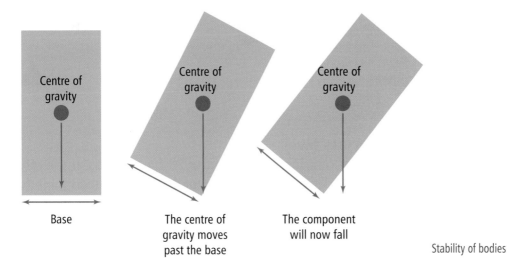

Stability of bodies

Stability

We can see in the first block that the centre of gravity is represented by a small blue circle. If a line was dropped vertically down from the centre of gravity it would be over the base. The base is the area that is touching the floor. As the block moves through an angle, the centre of gravity is moving over the base. If the centre of gravity goes outside the base then the block will fall. The block has become unstable.

(c) Look at the drawing on page 10: in order to make the stool more stable, three legs have been added. This extends the base and makes the stool stable.

Levers magnify forces

Levers are used to increase forces.

There are three main types, or classes, of lever:

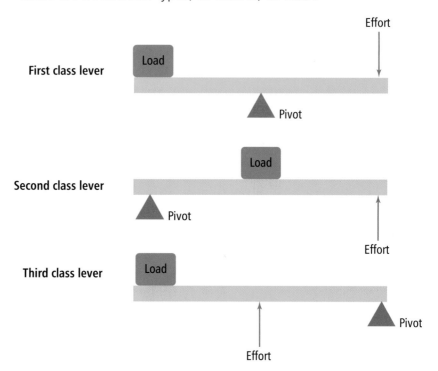

First-class lever

This type of lever can be used to lift heavy loads. If the effort is the same distance from the pivot as the load, then the effort and load are the same. If the lever is made longer the same effort will lift a heavier load but the effort will have to move further.

Second-class lever

A wheelbarrow is an example of a second-class lever. The pivot is at the end and the load is in the middle of the lever. When a wheelbarrow is lifted, the handles move a long distance upwards compared to the load being lifted. There is a small effort and a long distance moved at the handle, but a short distance moved and a heavier load lifted in the barrow itself.

Third-class lever

This lever has the load at one end and the pivot at the other end. The load is applied at the end of the lever. In this case there is a larger movement of the end of the lever than is applied at the load so this lever is not used to generate heavy forces. It is often used for gripping such as with tweezers.

Where can levers be found on a mountain bike? Make sketches of all the levers. What classes of lever are there?

Think
IT THROUGH

Forces

Forces act upon components and these forces can be classified into four types. Forces are measured in newtons (N).

On Earth the force of gravity makes objects accelerate towards the ground. The speed increases at a rate of 9.81 metres per second per second, which is written as m/s^2.

The force produced by a one kilogram mass on Earth is calculated as:

$$1 \text{ kilogram} \times 9.81 \text{ } (m/s^2) = 9.81 \text{ newtons}$$

As a rough guide, one newton is about one tenth of a kilogram (but this is only the case on Earth). A newton is not a very large force in engineering, and forces are often measured in thousands of newtons, or kilonewtons (kN).

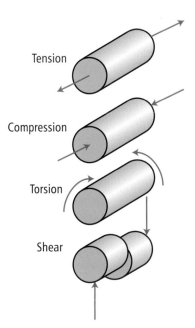

Types of force

Tensile force	This is when a material is being stretched
Compressive force	This is when a material is being squashed or compressed
Torsional force	This is when material is being twisted
Shear force	This is when a material is being cut or sheared

When components are subject to forces they can stretch, twist or deform. It is important to consider the strength of the material and the size of the components.

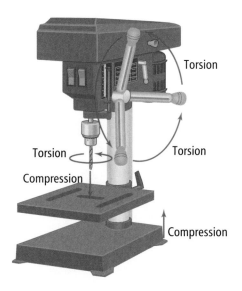

Types of forces on a pillar drill

Friction

Friction is a measure of how 'slippy' two surfaces are together.

We use the coefficient of friction, μ (the Greek letter mu) to measure friction, and it is based upon a scale of 0–1. The two materials are always stated when giving values of the coefficient of friction.

Sometimes components *need* to slide. Sometimes components need to almost stick together with friction. It is important to know how well you want components to slide when they are in contact with each other.

Here are some examples of coefficients of friction of different materials with steel:

μ	Material	Application
0.1	Lubricated metals	Bearings
0.2	Polythene	Light bearings
0.3	Brass	Bearings and locks
0.4	Cast iron	Machine beds and slide-ways
0.6	Tungsten-carbide	Cutting tools
0.8	Steel	Screw threads
1.0		

The coefficient cannot be zero, as it is impossible to have no friction.

When designers are thinking about what materials to use, they should always consider the effects of friction. Lubrication using oils drastically reduces friction. Bearings may have a coefficient of friction as low as $\mu = 0.03$.

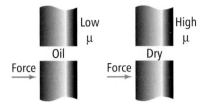

Bearing surfaces

Structures

Structures resist forces applied to them. A structure moves forces from one point to another. Structures are usually designed to withstand forces by redistributing them.

Structures can be:

- solid
- frame
- shell.

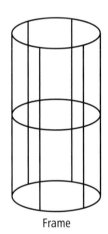
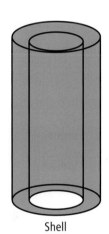

Solid   Frame   Shell

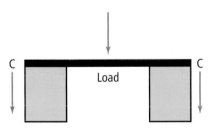
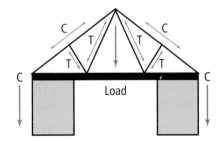

Tensile load = T
Compressive load = C

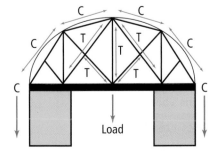
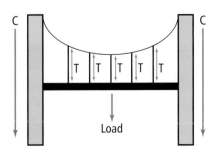

Framed structures

The bridge structures on the previous page show how frames can be used to redistribute forces. These types of structures can also be used in products.

Framed structures use less material, weigh less and cost less to produce than solid or shell structures. However, with framed structures there are often joints to be made.

Joining

There are many ways of joining components together.

Threaded fasteners

Screws and **bolts** are metal fasteners which have excellent strength. Both components to be joined are drilled and a bolt is placed in the holes. A nut is fixed to the bolt and tightened by threading. To apply enough force, a spanner is used. Threaded fasteners are reversible which means that they can be taken apart easily. Threaded fasteners are used where they may need to be removed, such as in machine covers and engine parts.

A threaded fastener

Rivets

Rivets are similar to bolts but are not threaded. They are placed in a hole then the end of the rivet is hammered over to form a dome or a flat surface. Rivets are not reversible. Once in place they cannot be removed without damaging them. Rivets are used in permanent joints, such as in the bodies of aircraft, ships and in bridges.

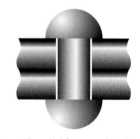

A non-threaded fastener (rivet)

Soldering and brazing

Metals can be joined by **non-fusion welding**. This process involves putting the two materials together and heating the metal to a specific temperature. When the materials are at the correct temperature, a **filler** is introduced. As the filler melts it fills the gap between the two materials.

THE JARGON DRAGON

filler – used to assist in the welding operation by providing extra material to join parts together

non-fusion welding – when the two pieces of metal to be welded do not melt. Only the filler material melts which holds the metals together

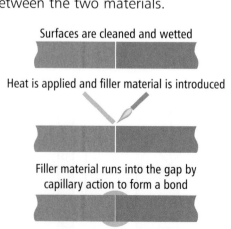

Surfaces are cleaned and wetted

Heat is applied and filler material is introduced

Filler material runs into the gap by capillary action to form a bond

Soldering is used for materials with a melting point temperature below 450°C and brazing for materials with a melting point above 450°C.

Welding can be used for steel and specialist materials with very high melting points. The joints made are permanent and waterproof.

Nails

Nails are used to join wood. They are not strong enough to join steel. They often support other means of joining.

Wood joints

There are several different methods of joining wood.

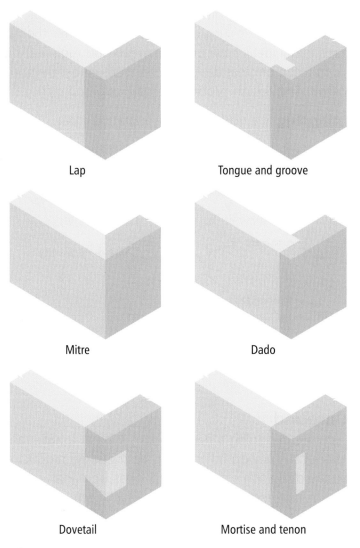

Lap	Tongue and groove
Mitre	Dado
Dovetail	Mortise and tenon

Adhesives

The most important skill in using adhesives is choosing the correct type.

Woods can be joined with adhesives such as PVA glue (polyvinyl acetate glue). Metal and ceramics require contact adhesives or epoxy resins. When you are joining two materials using an adhesive, the surfaces of both materials should first be cleaned.

Adhesives can be dangerous. You should always read the safety instructions on the packaging before using them.

Generation of ideas and solutions

The design specification gives all the guidelines for the solution to the problem, and an outline of what is to be achieved by the product.

mind mapping – a way of graphically representing ideas and linking them together

Some people find it very difficult to come up with ideas – this is quite normal. Good designs often come about after the designer's initial ideas have been changed and modified many times. Often the original design is changed almost beyond recognition. The design process is often long and difficult.

Mind mapping is a good way of generating ideas.

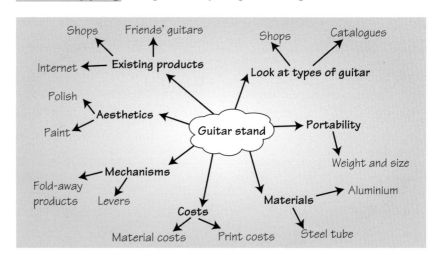

Example of a mind-mapping exercise to design a guitar stand

Look around

When you are designing a new product, look at a wide range of other products. They do not need to be the same product as the one you are designing – you can pull ideas from many different products. In fact, looking at products which seem at first sight to be very different can sometimes give you the best ideas.

Look at nature

Some of the most amazing structures can be found in nature – look close up at plants and spiders' webs. Birds' nests are fantastic structures, held together without fasteners or adhesives. Some animals have developed excellent defence techniques such as horns, shells, armour and long teeth. These are simple but effective designs. Some creatures change colour or shape. These are all good sources of ideas for designs of engineering products.

Sketch

Make notes and sketches of anything you think may help you with your design. Here are some quick sketches based upon the ideas from the mind map.

Look at the huge array of patterns, textures and colours found in nature too – you should find plenty there to inspire you!

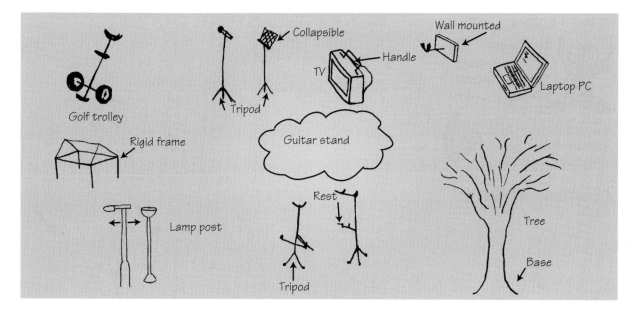

Notes are used to remind the designer of important features.

When generating ideas and creating mind maps remember:

- put down any ideas that come into your head
- do not leave ideas out because they do not sound very good
- the more ideas the better.

You can always eliminate the ideas you decide not to use.

Choosing from a range of designs

You will need to produce a range of designs from which to select the final design. This is better than producing only one design.

Many of the designs you produce will immediately be rejected because they are unsuitable. Eventually, all but one will be rejected until you have a single final product. For example, the numbers of ideas generated could be:

General ideas	10–25	Sketches and notes
Good ideas	6–10	Concept drawings
Short list of concept drawings	3–4	Detailed concept drawings
Final product	1	Full working drawings

Example: bracket design

A bracket is to be designed to hold a component to a machine. The designer sketches some basic shapes and ideas. As you have seen, designers need to have lots of ideas in order to select one final product.

These **concept drawings** are basic and show only the general idea of the proposed product.

This gives the designer a range of ideas to evaluate and choose from.

THE JARGON DRAGON

concept drawing – an initial sketch to convey design ideas and solutions

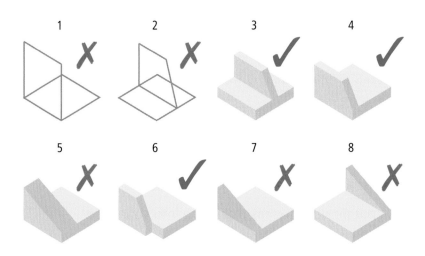

Concept drawings

Short list of proposed drawings

There should be about three good designs that could each be seen as a good solution. This is known as a **short list of design solutions**. The final three concepts are drawn in more detail and evaluated against the design specification.

At this stage the drawings are known as **concept designs**. They do not show the manufacturing details such as assembly methods, mechanism details, fastenings, finish or other information that will need to be considered before making the final product.

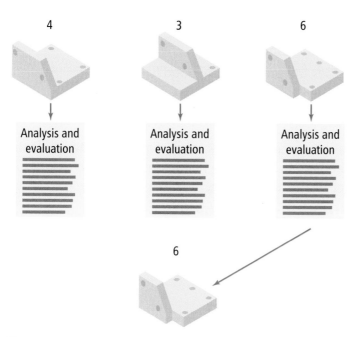

Details

When the concept designs are chosen these details need to be completed:

- How will the parts be joined?
- How will the mechanisms work?
- How much will the product cost?
- How much will the product weigh?

When new products are designed it is difficult to visualise what they will look like. Designers use different ways of showing what the new product will look like and how it will work:

- sketches
- perspective drawings
- orthographic projection
- isometric projection
- computer-generated images
- computer animation.

Evaluation

Evaluation is a way of constantly checking to see if what you have designed meets the design specification.

It is important to evaluate your work on an ongoing basis. Do not wait until the end as there are too many important things that may have changed.

Each time you evaluate you should consider:

- Will the product work?
- Does the product look good?
- Would potential customers buy the product?
- Are there better products on the market?

The design specification is broken down into the key features listed below. You therefore need constantly to check how your product relates to these key features.

Key features

- function
- quality standards
- styling aesthetics
- performance
- intended markets
- size
- maintenance
- production methods
- cost
- regulations

Constantly reviewing a design during its development is known as **value engineering**. The more detailed the design, the more detailed the evaluation.

Before production, a **prototype** should be built. This is an important stage at which to evaluate the product as it can be tested, seen and used by many different groups of people.

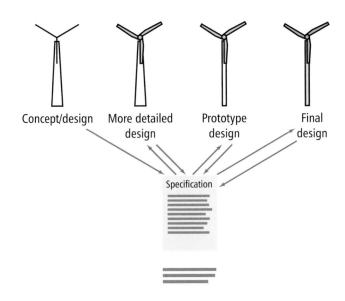

Concept/design   More detailed   Prototype   Final
design   design   design

Specification

Process of designing and comparing
with design specification

Imagine designing this wind turbine without evaluating the design!

Testing

Although designers can be very experienced at designing products, new products still need to be tested to make sure that they meet the specification. New materials or mechanisms may have been used that have never been tested before.

A series of tests will need to take place to ensure that each product performs properly when it is being used. Tests can be time consuming and expensive, so it is important to make sure that only the tests that are really needed are carried out.

Tests should aim to prove that the product meets the key features of the design specification.

Function

The product should be made to perform its task. It is sometimes important to test a product until it breaks to so that manufacturers know the limits of the product.

Completed wind turbines

Quality standards

The product should be tested to make sure it complies with all relevant standards. These could be British Standards from the British Standards Institution or European Standards. The International Organization for Standardization (ISO) is a network of national standards institutes from over a hundred countries working in partnership with international organisations, governments, industry, business and consumer representatives.

The British Standards Institution (BSI) was founded in 1901. Its role is to set standards for products. These standards ensure that a product is safe and practical.

Products that are to be sold in the European Union should be marked with the logo which shows that they have been manufactured to Central European standards.

Products that have been tested to British standards are able to carry the British Standard Kitemark logo.

These symbols show consumers which standards the product has achieved.

An example of British Standards in use

If a company's product meets a British Standard, the product is issued with a BS number and the company gets a certificate. The BS number is shown on the BSI Kitemark.

BS 5713 – this Standard is for manufacturers of sealed double glazing units and has been designed to cover the testing of the units as well as the manufacturing process. The units must pass tests for weathering and security of the seals. It is also compatible with European Standards.

British Standards can apply to almost any product, such as crash helmets, mountain bikes and skateboards. The system allows customers to make sure that the products they buy are fully tested and are of good quality.

Note that Britain operates in the European Union, and some British Standards (BS) and European Standards have harmonised. An example of this is **BS EN 71 Manufacture of toys**.

European Directives are sets of rules that all European countries must follow. The European Directive for toys sets out a list of requirements that toys must meet if they are to be sold in the European Union. If a product meets these standards then it can carry the CE mark. This is not a safety mark but it allows the product to be traded in the European Union.

BS EN 71 is broken down into six parts. These are not shown here in detail, but they give you an idea of the different things that need to be tested in order to meet European and British Standards:

1 mechanical and physical properties
2 flammability
3 migration of certain elements
4 experimental sets for chemical and other activities
5 chemical toy (toys that may have chemical in them but not a chemistry set)
6 graphical symbol for age warning label.

THE JARGON DRAGON

aesthetics – the appearance and form of a product or component

Styling aesthetics

A product will need to meet the styling **aesthetics** that were defined in the design specification.

Style aesthetics relate to the product's visual appearance. Style can be defined in terms of time, place, culture, people or anything else that influences how things look.

- **Time:** styles change through time, so a product could be designed, for example, in prehistoric style, stone age, middle ages, seventeenth century, 1960s, in fact any period of time in history or in the future.
- **Place:** style also varies with place, so the design could relate to a specific country such as Greece or Spain, a continent such as Africa, a type of landscape, a city such as Paris or New York or even another planet.
- **Culture:** style could relate to a type of religion, for example Buddhism or Judaism, a certain set of traditions or beliefs, such as Shaker, or a particular type of lifestyle, such as minimalism.
- **People:** Some people are associated with a particular style. For example, Victorian style is named after Queen Victoria. Here we see a Victorian-style lamp – typically, it is ornate with a curved pattern on it and was probably made by a blacksmith.

Styling aesthetics can be described in any terms that give an overall feel for how a product should look.

Style should always take into account a product's surroundings. This ensures that when the product is made it won't look out of place. When testing your product, therefore, you need to consider how well it will fit in with other products.

A Victorian-style lamp

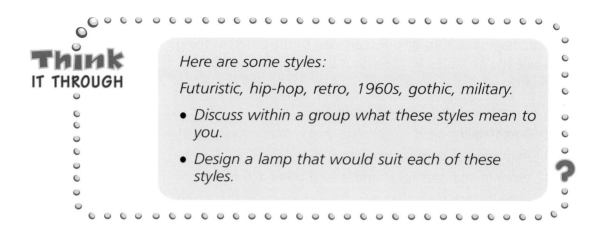

Think IT THROUGH

Here are some styles:

Futuristic, hip-hop, retro, 1960s, gothic, military.

- Discuss within a group what these styles mean to you.
- Design a lamp that would suit each of these styles.

Performance

Performance relates to how well the product can do the job it was designed for. Products can be tested for speed, strength, length of life, durability or any other characteristic the product needs to have.

Materials can be tested in a number of ways and the values for their basic properties can be obtained from the suppliers. However it is often difficult to predict how materials will perform when moulded into a new shape or used in a different environment.

Intended markets

The **market** is the group of consumers that the product is aimed at. Before products go into full production it is useful to make an initial batch to give to potential customers to try. The consumers can then give the company **feedback** about the product. If they are critical of certain aspects of the product, this will help designers to know what improvements to make.

Size

Products need to be tested for size as they may need to fit inside other products. They also need to be the correct size for packaging.

Maintenance

When a product is completed it will undergo **maintenance** checks to make sure that maintenance can be carried out according to the specification. It may be found that screws cannot be fastened, oil may leak out or the product may generally be difficult or time-consuming to maintain.

Production methods

The production method would be part of the **manufacturer's specification**. As a product is being made, the production method can be evaluated.

If, for example, a product is sand cast, this may be found to be too slow or inaccurate, so another method of production may be used instead, such as die casting. Unit 2 of this book gives details of various types of production methods.

Cost

A product's manufacturing cost will have been estimated before production began. The actual cost needs to be calculated and checked against the original estimate.

These costs relate only to the direct manufacture of the product. By the time the product finally gets to the shops, money will also have been spent on transport, insurance, warehousing and advertising which all add to the final cost of a product.

Regulations

Regulations are rules which must be followed. Rules devised in Europe are known as **EC Directives**. The government makes sure that the directives are met by developing **regulations** which businesses must follow.

Example regulation:
The Management of Health and Safety at Work Regulations 1999 (SI 1999/3242) (EC Directive 89/391/EEC)
These safety regulations are commonly applied in businesses. You may see signs or certificates displayed around companies' buildings to show that they work to these regulations:

- **COSHH:** Control of Substances Hazardous to Health
- **PUWER:** Provision and Use of Work Equipment Regulations.

The International Labour Organization (ILO) produces Codes of Practice which contain practical recommendations for ways companies should operate. Codes of Practice are not legally binding and do not replace national laws or regulations, or accepted standards.

The Health & Safety Executive (HSE) has produced Approved Codes of Practice (ACoPs) which businesses can work to.

Scale of production

As with all new products, the number of products required to meet market demand is only an estimate. It is possible that demand for a product will grow, which means that the scale of production will need to increase.

Small-scale production processes

These include making components by hand, such as cutting, filing, carving, stitching, manual assembly, manual welding and manual materials handling. Small-scale products are usually specialist products, and manufacturing time can vary from under an hour to a few months. Usually between one and fifty products are made at a time.

Medium-scale production processes

These include processes that use conventional machines such as lathes and milling machines, moulding techniques such as sand casting, and handling materials with a forklift truck. Typically 50–1000 products are made at a time, and production times for medium- as well as large-scale products vary from a few seconds per component to a number of days for complex products such as cars. Products may also be made in batches of, say, 10 or 20.

Large-scale production processes

Computer-aided manufacture, computer numerical control (CNC) lathes and milling machines, CNC burning machines, injection moulding, pressure die-casting, assembly using pneumatic devices and robots , and handling materials with automatic guided vehicles (AGV) are all examples of equipment and processes used during large-scale manufacture. From 1000 to millions of products can be made, usually by continuous production, such as a production line for cars.

The scale of production needs to be considered throughout the life of the product.

Modifications

When tests have been carried out, a lot of information is fed back to the designers. This allows the designers to make changes to a product's design. The drawings are modified, based upon the information obtained from the tests. Once all the tests are complete, the product is ready to go into production.

If further modifications are needed once the production process has started, this can hold up production. It is therefore very important to test products thoroughly.

Reasons why modifications take place during production

- New regulations might emerge relating to a product.
- Other components may change, which means making design changes to the product.
- Customers might complain about some aspect of the product, so it has to be changed.
- Safety issues.
- New materials and processes might be developed.

Sometimes cars that have been sold need to be brought back to the manufacturer for modification. This is know as a **recall**. Usually this happens when a fault has been found by some customers that could become a safety or maintenance problem.

Modifications are designed to improve a product, but sometimes problems arise because a product has been modified. There may be earlier versions still in stock that are now obsolete and can't be sold. Drawings need to be updated. These drawings need to be marked with a new modification number and date, and all previous drawings need to be destroyed.

2D and 3D drawing techniques

When designers are presenting their ideas to customers or managers, they need to consider the most appropriate method to do this. You will also need to select the best methods to show your designs. This section deals with choosing the best method of presentation for different audiences. The methods include:

- sketching
- shading (shading flat, shading cylinders, full shading)
- perspective and oblique drawings (single-point perspective, two-point perspective)
- isometric projection
- orthographic projection / working drawings
- assembly drawings

- modelling techniques (exploded drawings, dynamic presentations
- digital photography.

Sketching

Although there are many computer-based drawing packages about it is still very important to uses sketches. If you have a vision in your mind the best way to express it is by drawing it on some paper. Leonardo Da Vinci lived around 500 years ago and his sketches are famous for capturing his thoughts and visions.

A modern-day helicopter and Leonardo's helicopter sketch

Example sketches

Here we see a basic sketch of an air drill body. Sketches take a long time to get exactly right. Use construction lines to help with basic shapes.

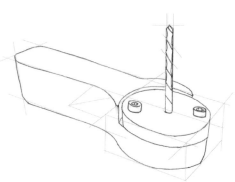

Basic construction sketches

Shading

Shading can bring a shape to life.

Shading cubes

Two-tone shading shows how light affects different sides of the block. Three-tone shading gives the top of the block a shade.

When drawing using a drawing software package it is possible to 'fill' areas. This reduces the need for pencil shading and gives a professional look to a drawing.

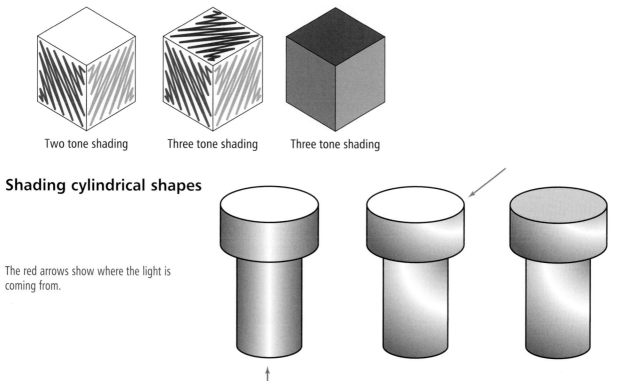

Two tone shading   Three tone shading   Three tone shading

Shading cylindrical shapes

The red arrows show where the light is coming from.

Shading drawings

Here we see simple shading. The light is the most important aspect of these drawings. Where the light hits the object there is no colour at all. These shapes are more difficult to shade than square objects, but shading can really show the object's form.

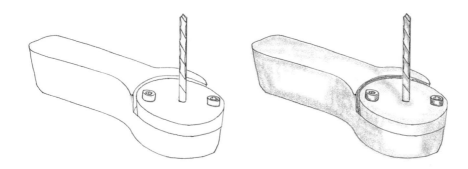

Example shading of air drill

The air drill has been sketched and shaded to give a real feeling of depth to the product. If you look closely you will see that the drill has a shadow.

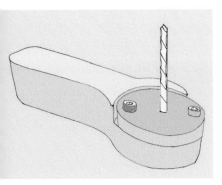

Using a fill facility on the graphics package 'Paint Shop Pro' the product is 'blocked out' with colour. The shading does not vary on the same plane but each plane has a different shade.

This can give a good, clear vision of what the product should look like. It is very easy to change colours, which is useful where plastics or painted surfaces are needed.

Perspective drawings

Perspective drawings are a type of 3-dimensional drawing. Below is an oblique projection and a perspective drawing. Let us compare the two types of drawing.

On the oblique projection, the lines that give the box depth are all at 30°, whereas on the perspective drawing they merge at a point in the distance known as the vanishing point. This gives a more realistic image than the oblique projection, but is much more difficult and time consuming to draw.

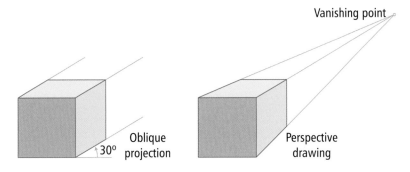

Below we can see how this applies to a more complicated product.

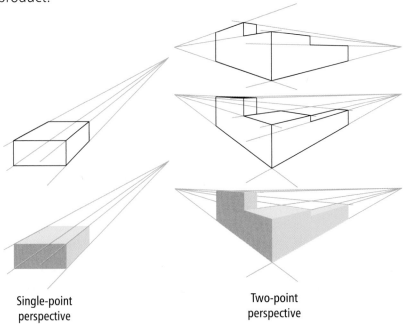

Types of perspective

As shown on the previous page, the lines can merge at one vanishing point, known as single-point perspective, or two points, known as two-point perspective.

Isometric projection

Isometric projection is a different method of drawing 3D shapes. It uses lines that are at 30° to the horizontal. Isometric drawings are not as realistic as perspective drawings, but they are very quick and easy to use.

Nearly all 3D engineering drawings are drawn using isometric projection.

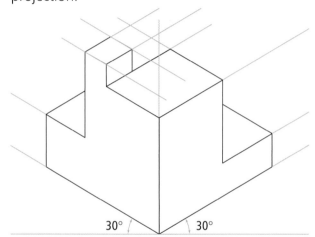
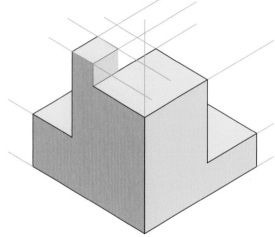

30°   30°

Orthographic projection

This is a formal method of drawing a product. It shows the product in two dimensions and from different views.

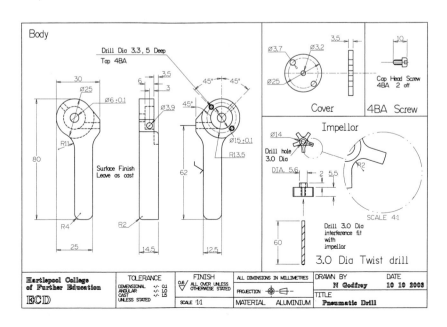

Orthographic projection example: air drill

Orthographic projections can include all the details necessary to produce the component.

The method uses a series of special symbols and notes and follows British Standard BS 8888. This Standard is widely used around the world, which means that engineers from all different countries can read these types of drawings.

Orthographic projections are also known as **working drawings** as they are the drawings that engineers use to make the components.

Assembly drawings

An assembly drawing shows the product with all the components in place. Most of the detail about each component is left off the drawing. The drawing is used to show how all the components go together or '**assemble**' and gives overall sizes. This is useful when designing the packaging or planning how the product will be transported.

In the assembly drawing, each part is labelled so that it is clear what components make up the product.

THE JARGON DRAGON

assemble– to put the individual parts of a product together

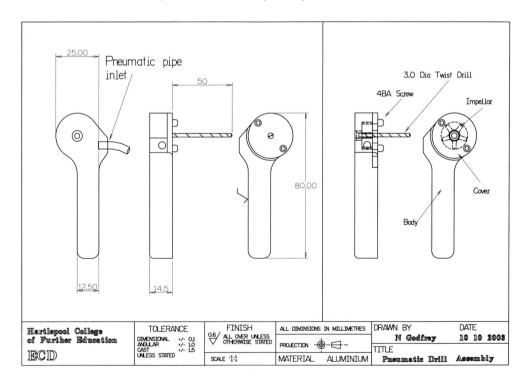

Models and prototypes

Before a product goes into production, manufacturers often make a model of the product. This may be a representation in card or wood, plastic or metal, or it may be a computer-generated model or image.

Models can show how mechanisms work. If the product is to form part of another product, the model can be put in its final position to make sure it fits, and that the complete product works.

Models can be used to show customers and clients what the finished product will be like, and how it will perform and look in the real world.

Modelling techniques

Three-dimensional computer models can be produced using software packages. Computer-generated 3D models can be rendered on-screen in any colour or texture.

The models can be rotated into any position, and you can also create exploded views.

Digital cameras

Digital cameras can produce images of models which can be superimposed into real-life situations. These images are used by sales and marketing departments for publicity, or to show clients how the product will look.

Digital images can be easily altered. In the images of the air drill, the handle length and the number of screws holding the cover in place have been changed.

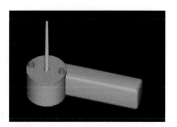

Air drill model produced using Studio Vis software package

Exploded views of air drill being rotated, created using Studio Vis software package

Digital photograph of the air drill being used (**Inset:** digital photograph of the product)

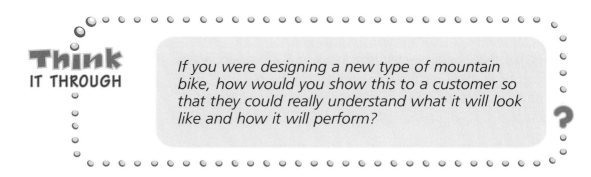

If you were designing a new type of mountain bike, how would you show this to a customer so that they could really understand what it will look like and how it will perform?

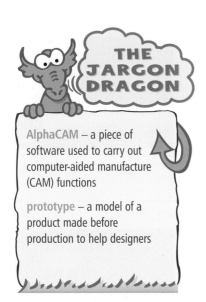

THE JARGON DRAGON

AlphaCAM – a piece of software used to carry out computer-aided manufacture (CAM) functions

prototype – a model of a product made before production to help designers

Prototypes

A **prototype** is a product that is not yet in full manufacturing production. It is used to determine whether there are any faults in the design or if there are any manufacturing problems.

Prototypes can be tested to determine a product's properties such as strength, and to measure the life expectancy of a product. Before full production goes ahead it is important that all these aspects are fully examined.

CAM (computer-aided manufacture) graphics

Animated graphics can show how a product will be produced by machines. The **AlphaCAM** system generates 3D graphics of computerised machining cycles for machines such as CNC machining centres.

Rendering

Using computer-based graphics packages it is possible to give surface textures and colours to graphical 3D models. These models can also be animated. The same technology that brings us films like *Toy Story* and *Shrek* is now being used to help design and sell products.

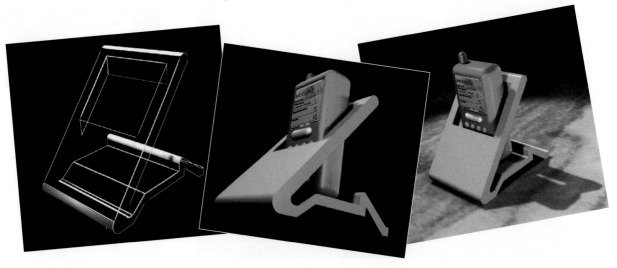

Communicating the design solution

Justification of your final choice

However many sketches, drawings or models you have produced or how many tests, notes or modifications you have made, you still need to make the final choice of what design of product to produce.

When deciding on the final product, remember always to ask yourself: Will it meet the agreed specification? The specification gives the guidelines for the design of the final product in the form of a list of its **key features** (see page 4). You therefore need to look at how the product relates to these key features.

There is no set formula for choosing the final design. There are simply too many elements to the design to make a simple formula work. However, there is a numeric method of evaluating a product, which is shown in the table below.

Example of a process of choosing a design

Key features		Product 1	Product 2	Product 3
Function	How well does the design work?	6	5	10
Quality standards	How well does it meet the quality standards?	4	5	10
Styling aesthetics	How well does it meet the style aesthetics?	5	5	5
Performance	What is the product's performance?	6	4	9
Intended markets	How well will it sell in the intended markets?	7	5	10
Size	How good is the size for the design?	9	5	2
Maintenance	How easily can it be maintained?	10	5	1
Production methods	How accessible are the production methods?	10	6	10
Cost	What is the cost compared to competitors'?	2	5	1
Regulations	Does it meet the regulations?	1	5	10
	Totals	**60**	**50**	**68**

Excellent: 7 to 10
Good: 4 to 7
Poor: 0 to 3

Each key feature is marked out of ten.

This method does have some disadvantages. You can see, for example, that Product 2 scored the least, with 50 points, but it was not poor in any key feature. This means you cannot depend completely on this method for working out your final choice.

In order to justify your final choice, you need to make sure that the product has all the key features listed in the specification.

Compare your choice with other products and highlight their weaknesses. For example, 'Product 1 is excellent in some areas but is poor in the area of meeting regulations and so cannot be picked'.

Referring to the design brief

You should also compare your product with the original client design brief.

Design brief	Product 1	Product 2	Product 3
How well does the product meet the design brief?	7	5	5
How well does the product meet the design specification?	3	6	5
Totals	**10**	**11**	**10**

Excellent: 7 to 10
Good: 4 to 7
Poor: 0 to 3

We now see that Product 2 has a high rating.

This list shows other elements of the design that could be measured:

Design requirements	Product 1	Product 2	Product 3
Physical and operational characteristics	5	4	10
Safety	5	5	10
Accuracy and reliability	6	4	9
Service life	7	5	8
Shelf life	8	6	6
Operating environment	5	5	7
Ergonomics	4	4	10
Size	5	5	2
Weight	6	4	3
Materials	6	5	2
Aesthetics	2	4	7
Production characteristics			
Quantity	3	7	5
Target product cost	4	8	6
Miscellaneous			
Customer	7	4	6
Competition	8	4	7
Totals	**81**	**74**	**98**

You can now see that, by taking into account design and production characteristics, some other issues have been highlighted. Further research can be used to collect opinions about the product:

- **Qualitative:** focus groups are small groups that give detailed feedback about the product.
- **Marketing:** ask the public, or the intended market, what they think of the product.
- **Quantitive:** obtain large quantities of data from potential customers through questionnaires.

As you can see, the whole process of choosing a final design is quite complicated, but you need to collect evidence in order to justify your final decision.

Details of your final design idea

Once you have chosen your final idea you must be able to communicate this to manufacturers. Use the most appropriate drawing techniques:

Working drawings:

- orthographic projection
- isometric projection
- types of fit
- types of finish
- notes and explanations.

Assembly drawings:

- exploded views
- how components move
- how components fit together.

The design brief is the document produced by the client to tell the designer what they want. You must be able to show the client that the final design fulfils the original design brief. This is done by:

- reviewing the design brief
- picking out all the important points
- checking how well the design matches the design brief
- commenting on how the product meets all of the important points.

case study

Air drill

Design brief

A vacuum-forming tool is a device that produces plastic components by laying a sheet of plastic over a cavity that has been machined into the shape of the product. Moulds vary in size and shape, depending on the component being made. Products range from baths to chocolate box trays.

During the production of vacuum-forming tools it is necessary to drill small holes into the surface of the mould itself. Mould cavities are made of aluminium so torque levels are low. The holes should be approximately 90° to the surface of the mould face.

The air is sucked out of the cavity by a pump, which causes the warm, soft plastic to flow into the cavity and to take the shape of the cavity. In this way, the component is formed.

You are to design a hand-held pneumatic drill that will drill holes into the vacuum-forming tool. It will be able to be rotated to any angle to suit the surface of the mould cavity. The product should have a low production cost and low maintenance. It should also meet the relevant manufacturing and safety standards.

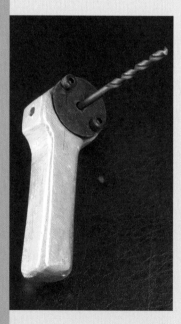

Reviewing the design brief

The design brief should be simplified by breaking it down into basic points:

- the device should easily be held in the hand
- the device should have low maintenance
- the device should have low production costs.

The designer should then break down each of these points further into more specific points. This helps the designer understand the design brief, and helps him or her explain to the client how the requirements have been met. Taking each point in turn:

Air drill

case study

The device should be easily held in the hand:

- lightness
- good ergonomics.

The device should have low maintenance:

- few parts
- simplicity in design
- easily maintained.

The device should have low production costs:

- low tooling cost
- low material costs
- ease of manufacture.

The device should meet manufacturing and safety standards for vibration:

- Standard Number: BS EN ISO 8662-7:1997.

Meeting the requirements

The device is manufactured from aluminium, which is a very light material. It has been ergonomically designed to allow good grip and a stable thumb position. Surfaces have a sand-cast finish for grip and all corner edges have a smooth radius.

There are only three main components. Screws are standard sizes and are easily available. The materials of all the major components are non-ferrous so that there can be no corrosion. The body is solid and robust and all inside corners are filleted to reduce stress.

The aluminium body is sand-cast, reducing the need for steel tooling. Aluminium is relatively expensive, but there is no waste through machining, and scrap can be re-melted. Casting is easy as aluminium melts at 660°C. The impellor is produced automatically on a CNC machine.

The product meets Standard Number BS EN ISO 8662-7:199. It has no moving parts (except for the twist drill).

As you can see, breaking down the design brief into specific points has helped the designer's explanation of how the design brief has been met.

Engineering drawings

Before exploring the types of drawing frequently used in engineering, it is worthwhile clarifying exactly why these types of drawings are so widely used. To do this, carry out the following activity.

Find a partner to assist you with this activity. It doesn't necessarily need to be fellow student – a friend or member of the family will also be able to help.

Find an object from around the house or classroom; preferably the object will include various features, shapes and dimensions – try not to pick something too simple or predictable.

1　*Be careful not to tell your partner what the product is and don't allow them to see it.*
2　*Provide your partner with a pencil and a piece of paper and sit back-to-back with the object in front of you.*
3　*Try to describe the product's features to your partner so he or she can make a sketch of what you are describing – allow around 3–5 minutes to complete the sketch.*

Have a look at a typical example below:

During an exercise like the one shown here, a student produced the sketch on the right from a verbal description

In general, students carrying out this exercise find the following common points:

- the drawing doesn't describe the product in sufficient detail
- the drawing isn't drawn to a scale nor does it contain any idea of size
- the object is usually drawn from one view only – often the easiest view that can be used to describe the product.

Unless the description is very clear the drawing partner will probably misinterpret several dimensions and features. In some cases they will even guess some of the features.

Students who achieve the best results tend to describe the product using very specific commands, such as 'Start with the bottom face – draw a line, approximately 10 cm long'.

The point of this activity is to demonstrate the most important reason for drawing to exact technical standards. This reason is **communication**.

THE JARGON DRAGON

communication – the verbal or non-verbal exchange of information

Communication is an important part of any activity and can be described as the passing on of information from one person to another. There are two types of communication:

- **Verbal communication:** talking to your friends and teachers, giving a presentation, giving instructions to a sports team-mate or working on a group activity.
- **Non-verbal communication:** writing an assignment, texting or e-mailing friends, drawing a picture, sign language or physical gestures such as waving or smiling.

Good communication is vital in everyday life and it is also an essential element of engineering. In engineering, ideas are often communicated using drawings.

Producing technical drawings is one of the most important aspects of any engineering process. They are used to pass technical information from one individual to another, or one company or department to another – usually involving a design and process stage.

To communicate effectively, engineering drawings must:

- contain all the information needed to carry out the activity
- be clear
- be of good quality
- be drawn to a recognised standard.

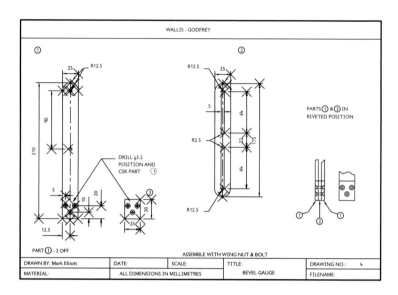

An example of an engineering drawing

Standardisation

If for example, different companies and departments used different methods and techniques to produce engineering drawings it would be very confusing for the various workers involved with manufacturing.

For this reason, drawings are carried out using a standard set of procedures, for example BS 8888:2000 (BS = British Standard).

Since 1927 engineers have produced engineering drawings to an exact set of standards used to ensure that all the people working in the industry can easily understand any engineering drawing.

The original standard, BS 308, was withdrawn in 2000 and replaced with the new standard BS 8888:2000.

This new version has been implemented to include adjustments by the International Organization for Standardization (ISO) and to take into account new developments such as CAD/CAM and 3D modelling (see Unit 3).

Some other important standards used commonly in engineering are:

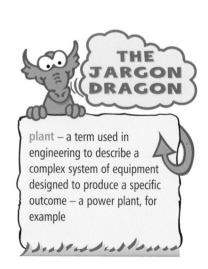

plant – a term used in engineering to describe a complex system of equipment designed to produce a specific outcome – a power plant, for example

- **BS 1553:** this specification is concerned with the standardisation of symbols for piping, heating, ventilation, power generation systems and compressing **plant**.
- **BS 2917:** this is the specification used to describe the graphical representation of symbols used for fluid power components, e.g. hydraulics and pneumatics.

- **BS 3939:** this specification refers to symbols used within the electronic, electrical and telecommunication sectors of engineering.

These are the standards that engineers must adhere to – the key features will be described later in this unit.

Getting started

The first part of any drawing activity is to select the paper size and orientation.

If you have printed a document from a computer program such as Microsoft Word you may have noticed the option to print in landscape or portrait mode. BS 8888:2000 also uses these terms to describe the orientation of the paper. The difference is shown opposite.

Also within Word you have the opportunity to select your preferred paper size. You will no doubt be aware that A4 is used as the standard paper size for printing most documents at school or college, such as your assignments etc. A3 paper is also popular and is commonly used for producing sketches – it is twice the size of A4. The 'A' before the number tells you that the paper is a standard size; the number is used to represent the actual size of the sheet.

'A' number	Size in mm
A4	210 × 297
A3	297 × 420
A2	420 × 594
A1	594 × 841
A0	841 × 1189

'A' series paper sizing

Title blocks and borders

Engineering drawings are required to have a border and a title block. A border (or frame) is used to determine the limits of the drawing area while the title block contains key information about the drawing, such as:

- the title of drawing – this could include the component title, process title or equipment description
- the name of the person who has produced the drawing (sometimes their initials are enough)
- the date the drawing was produced
- the drawing number and issue/revision number to ensure that the correct version of the drawing is being used
- the projection symbol (this is explained in more detail later on page 62)
- the scale of the drawing
- the materials specified.

In many cases a grid referencing system may be used to help in describing a particular feature on the drawing:

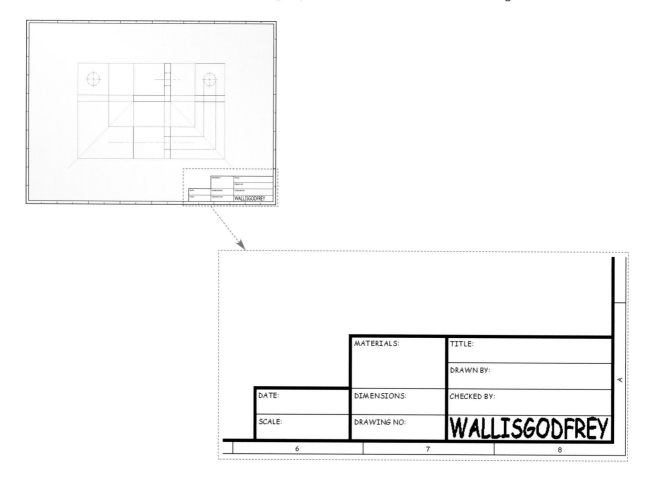

Drawing equipment

Exact manual drawing requires very specific equipment to produce the best results. The most important tools are described below.

The pencil

Arguably the most important tools used in design, pencils come in various standard grades:

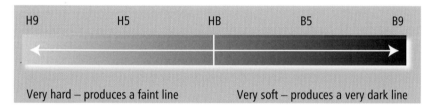

| H9 | H5 | HB | B5 | B9 |

Very hard – produces a faint line   Very soft – produces a very dark line

Mechanical pencils are now widely used, and are supplied based on the diameter of the graphite and the grade.

The eraser

Erasers are used by just about everyone to correct mistakes.

The rule

Again, the rule is a common tool used by students – usually to underline work.

In design, the rule is predominantly used to measure dimensions in both fractions of a metre (metric) and inches (imperial). The metric system is the standard choice for measurement in the UK.

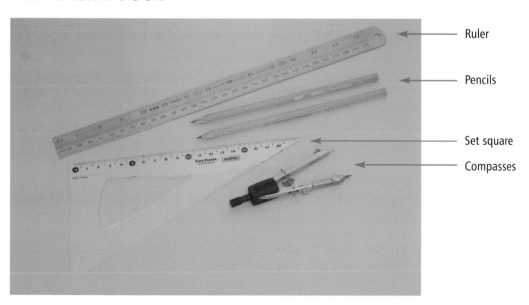

Ruler

Pencils

Set square

Compasses

The set square

The set square is commonly used by engineers to produce straight lines and edges.

It is used in conjunction with a T-square to produce lines at exactly 90°, 60°, 45° and 30° to each other. They can also be combined to produce multiple angles such as 75°.

The T-square

The T-square can be either found mounted to a traditional drawing board using pulleys (called **parallel motion squares**) or used freely by running a straight edge along one side of the drawing board.

It is used to produce straight horizontal lines and to provide a rest when using set squares. It is also normal practice for engineers to line up their drawing paper using a T-square as a reference plane.

The fine liner

A fine liner pen is sometimes used to embolden or highlight certain features on an engineering drawing. Similar to a felt-tip pen, a fine liner produces a very clean, consistent line.

The compass

Another well-used tool in engineering drawing is the compass. It is an easy, cheap and versatile method of producing different-sized circles and arcs.

Dividers

Dividers are used to transfer distance from a measuring device to the paper, or in some cases from one drawing to another.

They are similar in design to the compass, but dividers have two points instead of one.

Templates and stencils

Templates and stencils come in various shapes and sizes. They are used to produce standard symbols and text quickly.

Common types of templates are used to produce circles and polygons, and electrical, electronic, hydraulic and pneumatic symbols.

How to draw a border

You will need:

- *pencil*
- *rule*
- *masking tape*
- *flat table*

- *A3 paper*
- *T-square*
- *set square*
- *drawing board.*

Take a piece of A3 paper and set out your first engineering drawing, carefully following the stages below:

1 *Ensure your paper is as straight as possible, using your rule (or T-square if you are using a drawing board).*
2 *Using tape, secure your paper to the flat table or drawing board.*
3 *Draw in a border, carefully observing the dimensions shown below.*
4 *Draw in the title block, again using the exact dimensions.*
5 *Using your rule, place it about 1 mm above the title block lines and carefully write in the information required using block capital letters.*

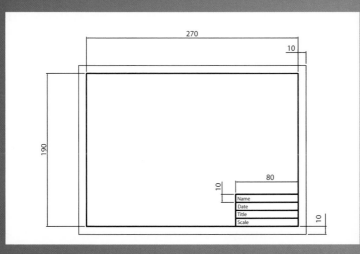

Note: It is best to measure all dimensions twice before committing to drawing construction lines and subsequent final emboldened lines.

Types of engineering drawing

Introduction

Engineering is a vast area encompassing millions of different products, equipment and services. Engineering can be divided up into different sectors and each of these vocational areas requires different types of drawings.

The main sectors of engineering are described in Unit 3.

Generally, drawings are used in various pieces of documentation, for example for maintenance and servicing, for producing customer specifications or for manufacturing/engineering requirements. The main areas are:

- concept generation of design briefs
- product specifications
- electrical and electronic wiring and circuits
- fabrication pattern development
- plumbing and pipework
- pneumatic and hydraulics circuits
- construction plans
- production processing plans.

To carry out the above requirements, the following types of drawing are used:

- sketches
- block diagrams
- flow diagrams
- perspective drawings
- orthographic drawings
- assembly drawings.
- exploded drawings
- schematic diagrams
- isometric drawings
- oblique drawings
- 3D models

Application of technology

All of the drawing types listed above can now be produced using various specialised computer-aided design (CAD) software packages. But whether you are drawing manually or using a CAD package, many of the rules and standards discussed in this section remain the same.

This section will focus mainly on manual techniques; however all the exercises have been designed to be compatible with CAD – for the best results try both manual and CAD drawing. (CAD is explained in more detail in Unit 3.)

Sketching

Sketching is an important part of technical drawing and is usually carried out for one of the following reasons:

- as part of the initial design stage to introduce and present an idea – sometimes called concept drawings

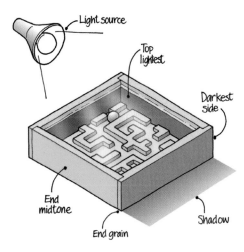

- to communicate an idea quickly from one source to another when there isn't necessarily time to produce a standardised drawing. This type of sketch may still contain several views and accurate dimensions.

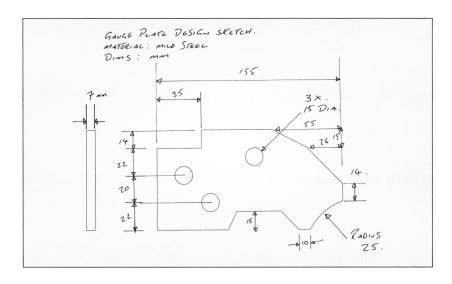

Unlike many technical drawings, there are no defining rules to producing a sketch and the standard will vary from one department or company to another – a CAD drawing could still be classed as a design sketch.

Design sketches can be very important to engineering companies and are often controlled and stored accordingly.

If concept designing:
- Pick a pictorial view that shows all the features of the product.
- Include shading to indicate a light source.
- Add colour but try to use only a few shades of one or two colours only (brightly coloured designs can sometimes look excessive and be confusing to the eye).
- Add text to highlight key features of your design – use capital letters and try to minimise the amount of information. (Large paragraphs of text can often be distracting, taking the viewer's eye attention away from the sketch, and may not be read in full. Keep to short, concise words or sentences.)
- Include a border and title block.

If producing a design sketch:
- Start by producing a border and title block. The title block should include all the relevant information about the drawing (see page 46).
- Use as many views as possible to ensure all the features are included. For example, a back view may contain a slot that is not obvious on other views. If some features are not on your sketch then problems processing the product at a later stage may arise.
- Draw clearly and neatly – it is not always necessary to use a rule or compass but use these tools if required.
- It is not essential that you draw to an exact scale, although it helps to visualise the product if you can provide a general idea of proportion.
- Include all essential dimensions including **chamfers**, **fillets**, radii and diameters.

Block diagrams

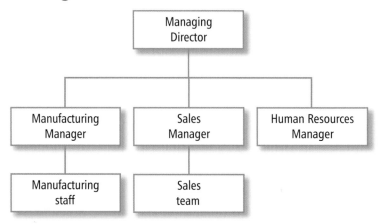

Block diagrams are often used to demonstrate the connection between elements of an activity or process.

They may be used to describe the components of a manufacturing process, a company staff chart, a sequence of management events or used as a quality control tool.

Flow diagrams

Flow diagrams were first used by computer programmers in the 1960s and have since become popular in many engineering and business activities.

These types of diagrams are used to document a process or activity for analysis or problem solving – they often aid in the understanding of a process.

Flow diagrams allow the person reading them to 'walk through' the process without the need to physically see it. This type of analysis can often uncover potential problems, solve process problems, find bottlenecks, and reduce unnecessary activity steps.

When producing block and flow diagrams:

1 Ensure the work is clear, well spaced out and neatly produced.
2 Keep the blocks and symbols a uniform size – if possible use a stencil.
3 Write notation in capital letters.
4 Keep arrows at angles of 90° with clear, well-defined arrowheads.

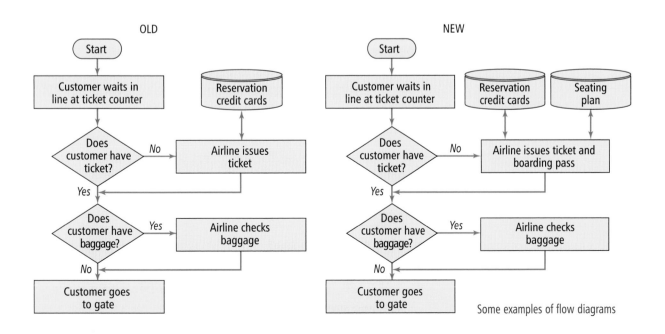

Some examples of flow diagrams

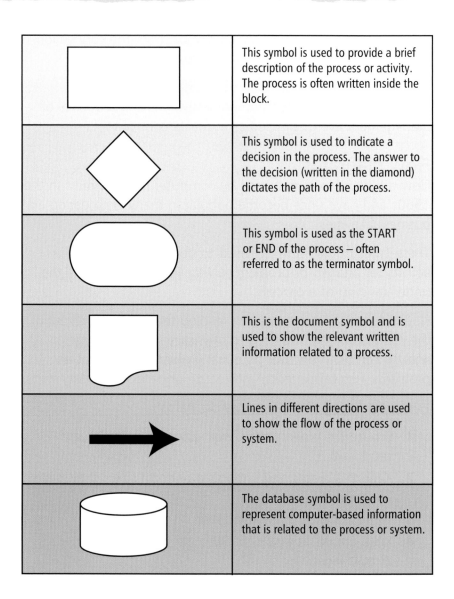

	This symbol is used to provide a brief description of the process or activity. The process is often written inside the block.
	This symbol is used to indicate a decision in the process. The answer to the decision (written in the diamond) dictates the path of the process.
	This symbol is used as the START or END of the process – often referred to as the terminator symbol.
	This is the document symbol and is used to show the relevant written information related to a process.
	Lines in different directions are used to show the flow of the process or system.
	The database symbol is used to represent computer-based information that is related to the process or system.

Symbols used in flow diagrams (all of these symbols and more can be found in MS Word within Autoshapes – Flowchart)

Perspective drawing

Perspective drawing is not widely used when producing technical drawings – it tends to be used more often by architects, artists and graphic designers when drawing objects or landscapes, usually when there is a need to convey space and distance.

It is very useful, however, for engineers to understand perspective drawing, particularly in relation to sketching. Perspective drawings have several advantages over technical drawings:

- they are easy to read
- they are quick and cheap to produce
- they make the subject seem more real because it is the way we see things naturally.

Perspective drawing works by focusing on a vanishing point within the drawing that is used as a reference point. The vanishing point is simply used to show the point where objects become 'out-of-sight'.

The most common types of perspective drawing are one-point and two-point perspective.

The diagram below shows a one-point perspective drawing of three simple blocks. Note that one of the principal faces is parallel to the picture frame.

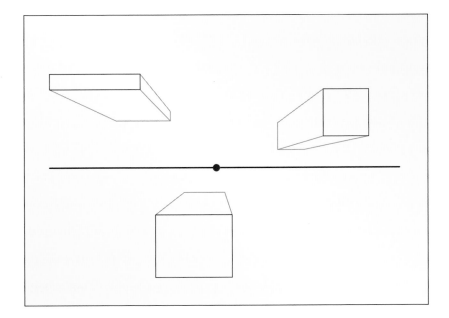

To sketch this out, follow the steps below:

1 Draw in a horizontal line to represent your horizon.
2 Pick a vanishing point on that line (the blue dot on the first diagram).
3 Draw a square anywhere on the page; note:
    a if you draw the square above the horizon, the block will appear to be above the viewer
    b if you draw the square on the line it will appear to be at the same height (the exercise will not work if the square is drawn directly over the vanishing point)
    c if you draw the square below the line, the finished object will appear to be below the viewer.
4 Draw a line from the corners of the square to the vanishing point.

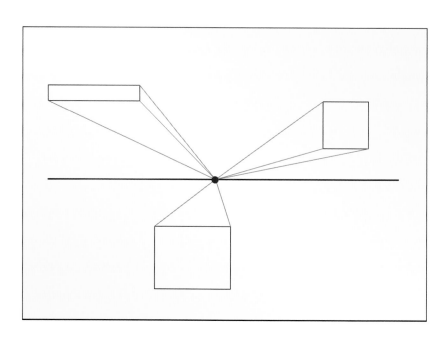

5 To finish, simply draw in the vertical and horizontal lines to convey the depth of the object.

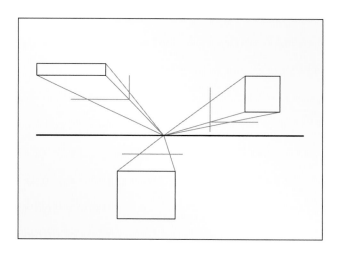

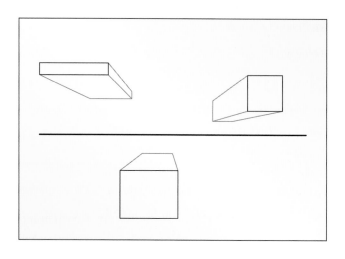

The drawing below shows a two-point perspective drawing of two blocks.

This type of perspective is commonly used for engineering sketching; note that this drawing shows the cubes from an inclined view:

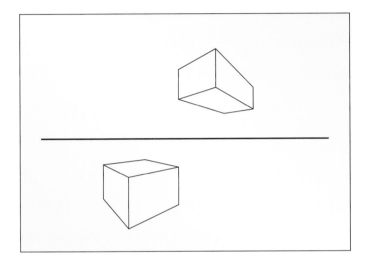

Many of the steps are very similar to those for one-point perspective drawing:

1   Draw in a horizontal line to represent your horizon.
2   Pick two distant vanishing points on this line.
3   Draw a vertical line of any height somewhere between these points – the height of this line will represent the height of your object at its nearest to the viewer:
4   Draw construction lines from each end of the vertical line to the two vanishing points.

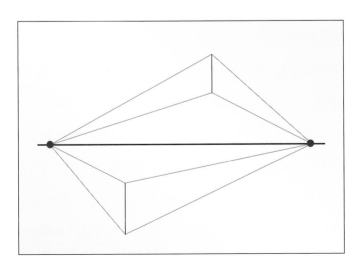

5   Draw in two further vertical lines on each side of the first vertical (keep between the construction lines). This is to convey the depth on each side and they don't necessarily need to be equal. If your object is on the horizon line then the object will be finished at this point.

6   If the object is above the horizon draw a construction line from the bottom of the two new vertical lines to the vanishing points. If the object is below the horizon draw a construction line from the top of the two new vertical lines to the vanishing points.

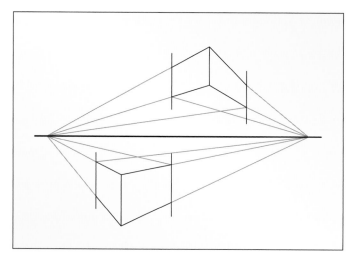

7   To finish, simply draw in the lines a little bolder to see the limits of your object.

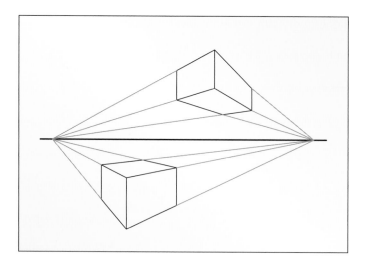

When drawing in perspective, if you are unsure where to draw the object with respect to the horizon, remember the following:

- small-scale objects look best just below the horizon
- human-scale objects look best on or just beneath the horizon
- large-objects tend to lie in the middle of the horizon.

Search for images of 'perspective drawings' on the Web. Notice how they all follow the same general rules as the ones above (the artists have simply had a little more practice than you).

Orthographic projection

Orthographic projection is the most widely used method of producing technical drawings for engineering products.

It is a standard method of representing a product so that all features (lines, edges, holes, radii) are included in the drawing. It is also the preferred specification document used by many engineers to actually manufacture or fabricate a product.

Orthographic projection can sound a little daunting at first. However once introduced it is a relatively straightforward way of representing objects.

It involves drawing an object or product using different views which are situated at right angles (90°) to each other.

This type of drawing does not produce a 3D image (as does pictorial drawing such as isometric or oblique), thus it requires the viewer to interpret the individual views to get a clear idea of what the product looks like.

When objects become more intricate, orthographic drawings inevitably become more complex – although with the appropriate training and practice they are not difficult to produce.

There are two forms of orthographic projection:

- **first angle** is used in the UK and Europe
- **third angle** is used in the US.

Views of orthographic projection

The easiest way to understand orthographic projection is to ensure that you first understand the different views.

The photograph at the top of the next page shows a well-known product.

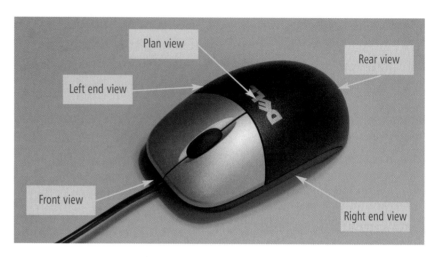

The photo includes some text, which represents the principal views (note how they are all at 90° to each other):

- plan view (normally looking from the top of the object)
- front view (often the view providing the most detail)
- right-end view (from the right-hand side)
- left-end view (from the left-hand side)
- rear view (from the back of the object).

There are no set rules when producing a drawing as to how many views should be used. In practice, normally three or four are sufficient: plan view, end view (left or right) and front view.

If the photo was set out using first angle, it would look like this:

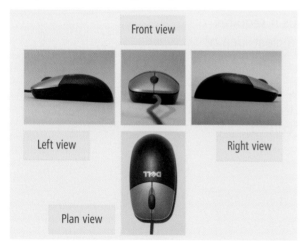

If the photo was laid out in third angle, it would look like this:

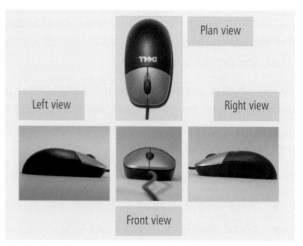

Note that the only difference between the two projection methods is the layout of views.

Now we have clarified the different views used by orthographic drawing. We now need to apply this to a 3D drawing of an object. At the top of the next page is a pictorial drawing of a simple bearing.

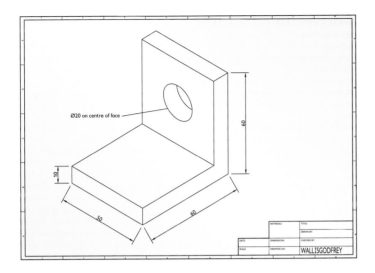

A simple bearing

Drawn in first angle it would look like this:

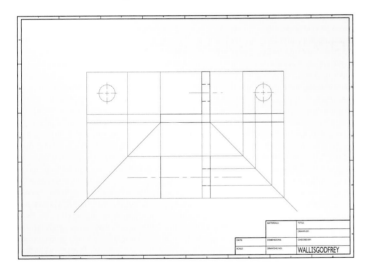

Drawn in third angle it would look like this:

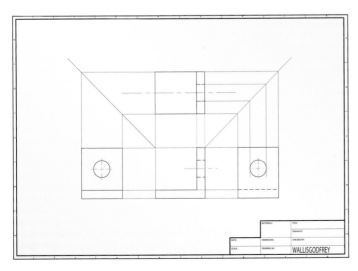

Symbols of projection method

Both first- and third-angle projection use a standard symbol to show which method is being used.

First-angle projection symbol   Third-angle projection symbol

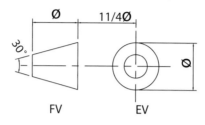
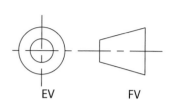

FV   EV   EV   FV

To draw these symbols accurately use the dimensions shown.

Note: ∅ represents the circle diameter and can be drawn at any sensible size.

How to draw using orthographic projection

As explained above, first- and third-angle projection show exactly the same views just in a different format – as first angle is more commonly used in the UK we'll look at this method in more detail.

You will need:

- A3 paper
- pencils (2B and HB)
- rule
- set square
- drawing board with T-square
- eraser.

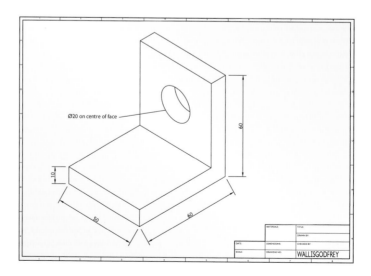

As usual, start with your border and title block – this time you can draw in the projection symbol in one of the spaces.

Divide the paper up into three areas by drawing in two vertical construction lines (use a harder pencil so that these lines are feint and can be easily erased later on) – these lines will represent the length of the object.

Then draw in two horizontal construction lines – the first should be placed just above the centre of the paper and the second will represent the height of the object.

1 Using the correct equipment draw in the front view using the dimensions shown – draw in the construction lines first with feint lines so that mistakes can be easily erased.

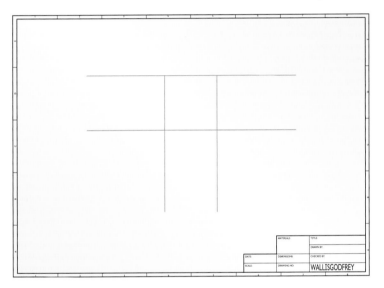

2 Once the front view is constructed to the required standard (check with your teacher if you are unsure), draw in vertical lines down from all the key sides, edges and features:

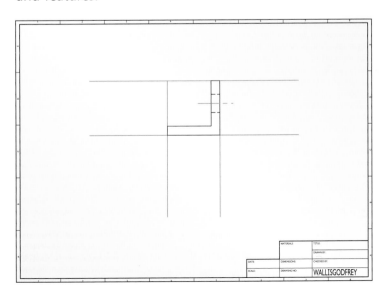

Leaving a sensible space (see note below) draw in a box to represent the dimensions of the plan view – don't forget, this is looking down on the object.

Note: To ensure that you leave the correct space between the plan and front view, you will need to remember:

- you may be required to draw in dimensions or notes in the space
- if the drawing is too cramped, it will appear cluttered and possibly confusing to the viewer
- if neatly spaced out, the drawing will be much more accessible and easy to understand
- most importantly – the space between the front and plan views will need to measure exactly the same as the distance between the front and edge views
- if you do not consider the spacing properly, all the views might not fit on the same piece of drawing paper.

3 Draw in construction lines horizontally across from the right of the front view to represent the key features.

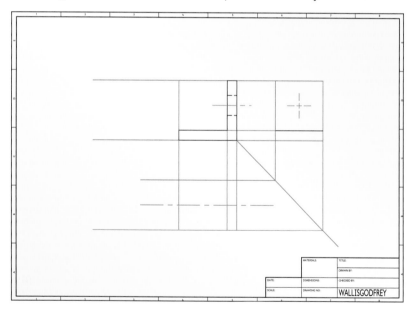

From the bottom right-hand corner of the front view, draw in a line of exactly 45° as shown above (you will need to use the correct set-square).

4 From the plan view draw construction lines to intersect the 45° line. At the intersection point draw a line vertically up – this will represent the features of the plan view as projected onto the side view:

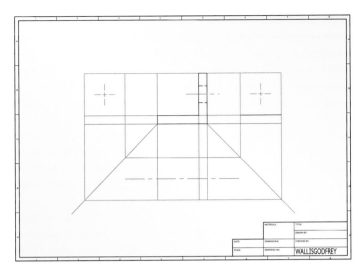

Using a softer pencil, embolden all the key sides, edges and features.

5 Erase all construction lines.

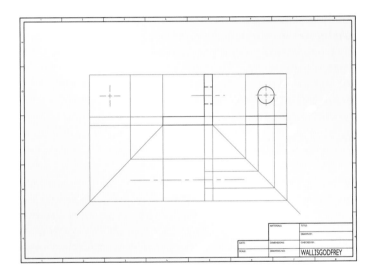

6

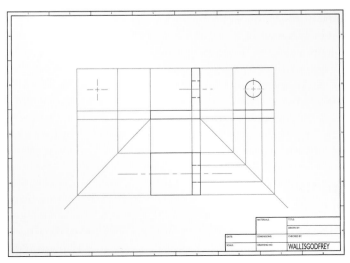

7

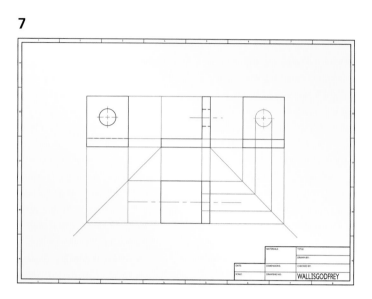

To draw the object using third-angle projection, note the following differences:

- Start with a horizontal line about 20 per cent of the way up from the bottom of the page – draw your front view here.
- Project upwards and draw in your plan view.
- To draw the end view extend a line from the top-left of the object and repeat as described previously.

Line type

One of the most important features of an orthographic drawing is the type and thickness of line used. This is not decided by the individual designer – the line type is set out in BS 8888:2000.

Line type and thickness are used to convey very specific information to the reader, such as whether the line shows the outline or edge of an object, a line that is hidden from view, a centre lines of holes, etc.

If the designer chooses not to use the standard types then the finished drawing could lack clarity, resulting in the viewer (for whom the drawing was intended) being confused.

The simplified version on the next page shows the different types of line commonly used:

Dimensioning

Dimensioning is vitally important when producing engineering drawings to show the exact size of features contained within the drawing.

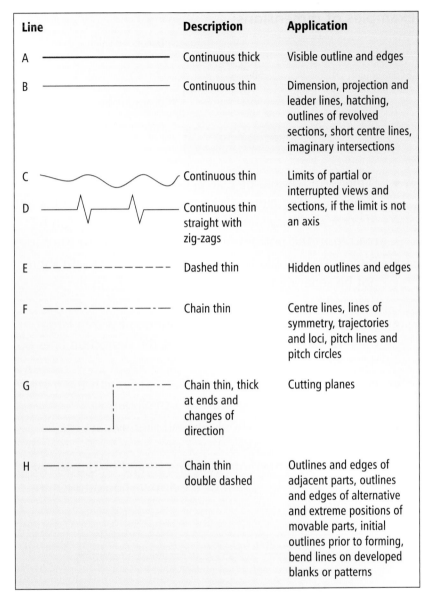

Line	Description	Application
A	Continuous thick	Visible outline and edges
B	Continuous thin	Dimension, projection and leader lines, hatching, outlines of revolved sections, short centre lines, imaginary intersections
C	Continuous thin	Limits of partial or interrupted views and sections, if the limit is not an axis
D	Continuous thin straight with zig-zags	
E	Dashed thin	Hidden outlines and edges
F	Chain thin	Centre lines, lines of symmetry, trajectories and loci, pitch lines and pitch circles
G	Chain thin, thick at ends and changes of direction	Cutting planes
H	Chain thin double dashed	Outlines and edges of adjacent parts, outlines and edges of alternative and extreme positions of movable parts, initial outlines prior to forming, bend lines on developed blanks or patterns

Types of line

Without adequate dimensioning, processing of the product would become difficult, as engineers may find the drawing unclear, leading to mistakes and guesswork.

Dimensions should be clear, accurate, well spaced and follow these guidelines:

- Each dimension required to define the object needs to be shown only once on the drawing.
- There should be only the dimensions necessary to understand the object – extra dimensions with no value should not be included.
- Linear (straight) dimensions should be expressed in millimetres, angular dimensions should be expressed in degrees.

Examples of dimensions

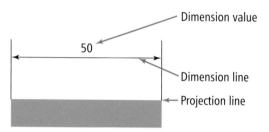

- Dimensions should be placed in the middle of a dimension line, either above it (if horizontal) or to the left (if vertical) – they should be placed so that they can be read from the bottom or the right-hand side of the page.
- Dimensions should be placed off the drawing and should not be separated or crossed by other lines on the drawing.
- Smaller dimensions should be positioned within larger dimensions – this is to ensure that the projection lines do not cross.
- Projection lines should line up accurately with the feature they are dimensioning, leaving a small gap.
- The projection lines should extend slightly past the dimension line – usually 1 or 2 mm.
- The arrows on the dimension line should be drawn neatly and uniformly. The point of the arrow should touch the projection line.

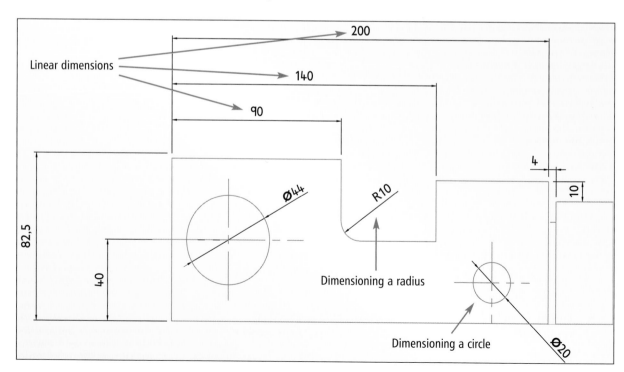

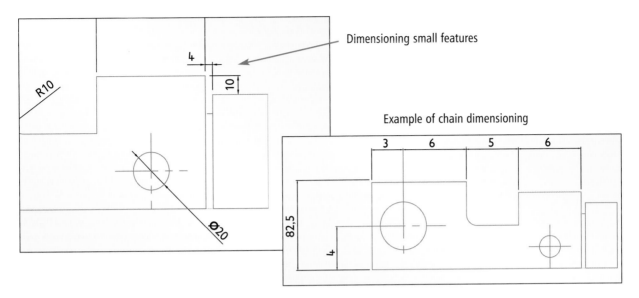

Dimensioning small features

Example of chain dimensioning

Standard components

When you are producing an orthographic drawing it can prove time consuming and tedious to draw many standard components such as screw threads, bolts, gears and springs.

If unsure why, just think about drawing in accurately all the features of a screw or bolt thread or all the teeth on a bicycle spur gear – it would take ages!

To get around this problem, BS 8888:2000 recommends the use of certain **conventions** to represent such standard components.

THE JARGON DRAGON

convention – an agreed method of representing a feature

Shown below are some common conventions used in technical drawing:

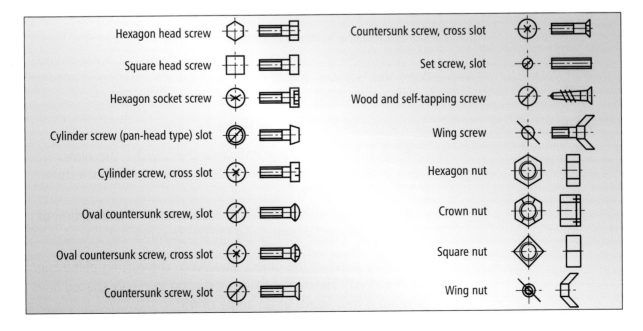

When producing engineering drawings you will often need to clarify complex products, which may have a large number of features.

Similarly, it is sometimes useful to be able to simplify a drawing by reducing the repetitive information required to make up an object view.

Partial views

If it is necessary to include more detail about a feature, sometimes a partial view is used. This view would include all the additional information that the original view might not have space to display.

In many cases it is useful to draw an enlarged view so that the viewer can better understand the features of the product.

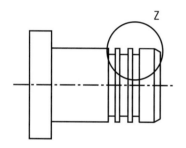

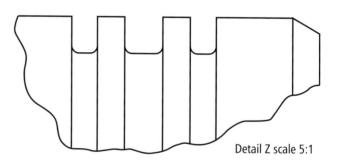

Detail Z scale 5:1

Sectional views and cutting planes

Sectional views are used when an object has important features that are not apparent when looking at the external view of it.

An easy way to think about this is using a birthday cake. From the outside all you can see is icing, but if you cut away a piece, then this 'section' would contain all the information about the inside of the cake – the chocolate, jam, raisins, etc.

Sectioning works in exactly the same way:

- A cutting plane line intersects the object – the view required is clearly represented using two capital letters with an arrow.
- This view is then drawn in separately (in the direction of the arrows) showing the hidden features.

Note: Any parts of the object 'touching' the cutting plane line are 'hatched'. Hatching involves drawing in evenly spaced lines (around 4 mm apart) at an angle of 45°.

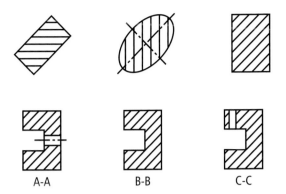

A-A   B-B   C-C

There are other common methods of simplifying drawings to make the viewer's interpretation of them easier. These include the following.

Repetitive features

Repetitive features are used if the product has a high number of exact features such as multiple drill holes – this prevents the designer carrying out needless repetition.

Only one of the features is drawn in full detail – the rest are represented by centre lines.

a) Holes on a circular pitch

Interrupted

Interrupted drawings are used if the product has a long, uniform feature such as a shaft – it is normally used to save space on the drawing.

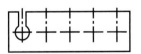

b) Slots on a linear pitch

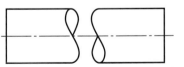

a) Conventional break lines for solid shaft

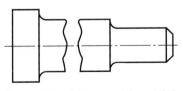

d) Type C break lines used for solid shaft

b) Conventional break lines for hollow shaft

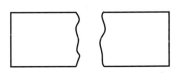

c) General break lines (type C lines)

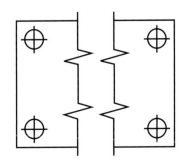

e) Type D break lines

Assembly drawing

Assembly drawings are sometimes called **general assembly drawings**.

This type of drawing is used to show a complete finished product with all the parts assembled into their correct positions.

Assembly drawings tend to be used for complex products with many different parts, such as a car engine.

An assembly drawing may include overall dimensions with accompanying notes such as fitting instructions.

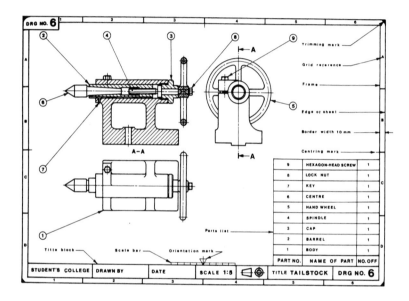

Adapted from Ostrowsky, O (1989) *Engineering drawing: with CAD applications.* Edward Arnold

Features of an assembly drawing

The assembly drawing will usually contain all the normal features of an engineering drawing, i.e.

- a border and reference grid
- a title block
- a drawing of the component in orthographic projection
- appropriate dimensions
- surface finish requirements
- manufacturing, fitting and assembly instructions.

In addition to these (most of which have been covered earlier in this section), the assembly drawing may contain a parts list.

The parts list

This is included on the assembly drawing when many different components are used to make up the final product.

An individual component within the assembly is identified using an **arrow (leader) line** and **reference notation** – this is normally a number contained within a circle or 'balloon'.

Using the reference notation on the diagram, the title block is used to describe the part in more detail. Typical information might include:

- the part name or reference number
- a description of the part
- the material used to make the part
- the quantity required
- a grid reference.

Exploded drawings

Exploded drawings are pictorial representations of all the component parts used within an assembly.

They are often used in industry:

- sales and marketing staff use them to promote or sell a product
- maintenance teams use this type of drawing in equipment manuals when carrying out the service and repair of machinery
- manufacturing personnel (who may not have been trained to read technical orthographic drawings) use exploded drawings to carry out assembly work on production lines.

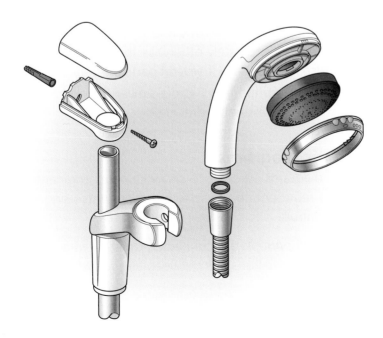

Exploded diagrams can also be found in many handbooks for products you find around the home, for example:

- hand-held power tools
- lawnmowers
- flat-pack furniture kits
- model-making kits
- car manuals
- vacuum cleaners
- instruction manuals for computer equipment and gaming consoles.

Tips for making an exploded drawing

Exploded drawings tend to be the most difficult to produce and as a product becomes more complex (with many components), so does the drawing.

Exploded drawings are drawn pictorially, normally in an isometric view so that the viewer can get a clear idea of the orientation of the product. This orientation is extremely important so that the different components can be located easily on the final assembly.

Here are some important tips when making exploded drawings:

- Use an isometric scale to draw the exploded view.
- Arrange the parts in the correct sequence, i.e. the order in which they would physically need to be assembled.
- Space out the components so that each one can be seen clearly and is distinct from the ones next to it.
- Draw the scale, dimensions and features of the individual components accurately otherwise the viewer may become confused – especially if several parts are very similar.
- Label each component – use a similar method to that described for assembly drawings.

Schematic diagrams

Schematic diagrams are used to show complex engineering activities by representing the system components using standard symbols.

These standard symbols are connected together in a 'circuit' to show clearly the relationships between all the system components.

The common types of schematic diagrams are described below:

- electrical wiring diagrams
- electronic circuit diagrams
- pneumatic circuit diagrams (fluid power systems)
- hydraulic circuit diagrams (fluid power systems)
- plumbing circuit diagrams
- environmental circuits such as heating, air-conditioning and refrigeration
- piping diagrams.

As described earlier, standardisation in the form of British Standards such as BS 2917 (fluid power systems) and BS 3939 (electrical, electronic, telecommunications) dictate how these diagrams are presented and used.

An example of each one is shown below.

Electrical wiring diagrams

These types of drawings are used by electricians and maintenance teams when installing or repairing electrical mains.

For example, your house will contain various **ring mains** (which connect the sockets together) and **lighting circuits** (which include fittings and switches). When the house was being built, an electrician would have used an electrical wiring diagram to determine where the sockets, lights and switches should be placed.

In industry electricians use a similar type of drawing to install equipment such as conveyors and robotics, in addition to the more regular tasks of lighting and plug sockets.

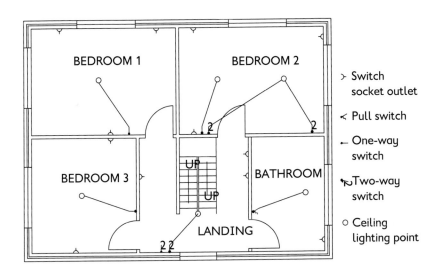

Electronic circuit diagrams

Many products now contain electronic components such as resistors, diodes and transformers – electronic circuit diagrams are used by engineers and designers to determine the requirements when setting out an electronic circuit.

These diagrams are used to show the reader how the components are set out and connected on a printed circuit board (PCB). Specifically, they show the reader how the components (represented by standard symbols) are connected together (usually by using a copper track).

The symbols used for this type of drawing are found in BS 3939; an example is shown below:

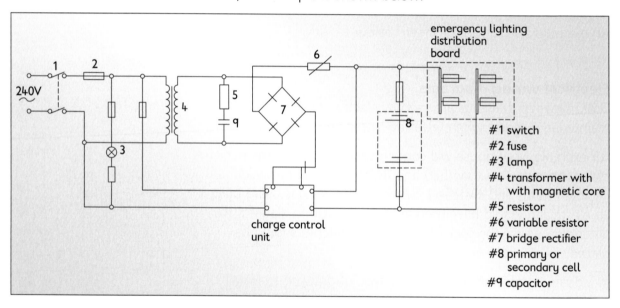

#1 switch
#2 fuse
#3 lamp
#4 transformer with with magnetic core
#5 resistor
#6 variable resistor
#7 bridge rectifier
#8 primary or secondary cell
#9 capacitor

Fluid power schematic diagrams (hydraulic and pneumatic diagrams)

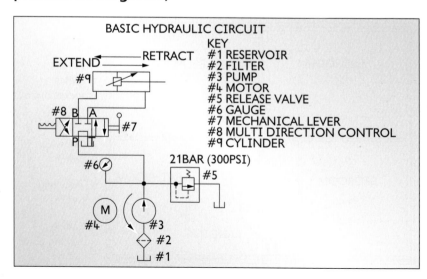

BASIC HYDRAULIC CIRCUIT
KEY
#1 RESERVOIR
#2 FILTER
#3 PUMP
#4 MOTOR
#5 RELEASE VALVE
#6 GAUGE
#7 MECHANICAL LEVER
#8 MULTI DIRECTION CONTROL
#9 CYLINDER

Fluid power diagrams are used to represent circuits powered by **pneumatics** (compressed air) and **hydraulics** (oil pressure).

The symbols used represent the components in these types of drawing. They can be hard to understand and will not be discussed in detail here. The standard symbols are described within BS 2917.

Plumbing, environmental and piping diagrams

These types of drawing tend to be used widely in the construction industry to install and maintain various systems.

These systems include domestic and industrial water supplies; fittings and waste pipes; industrial water and steam piping; and domestic and industrial heating systems. An example is shown below.

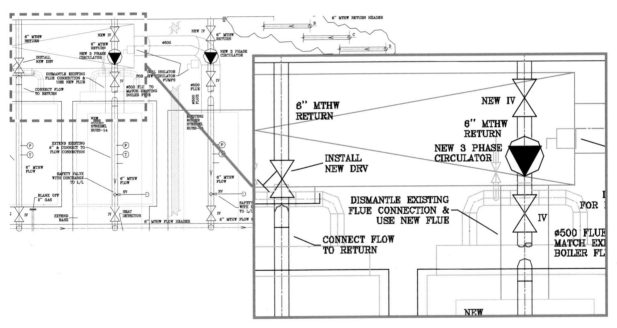

Tips for drawing schematic diagrams

- Plan your drawing carefully to ensure that all the component symbols are spaced out neatly, leaving enough space between them for the connection lines.

- Sketch out the diagram on scrap paper first if necessary.

- Draw in a border and title block, as described earlier.

- Ensure that you keep all the symbols to a uniform size, if possible using a stencil.

- Draw in the network of connection lines used to represent the electrical wiring, PCB copper track or piping.

- Once you are happy with the spacing draw in the standard symbols in the correct positions on the lines.

- Include any writing or notes using uniform block capital letters.

Pictorial drawings

Pictorial drawings are accurate technical drawings, which show a component in 3D as opposed to 2D – like orthographic projection, for example.

The advantage of a pictorial view over an orthographic projection is that it is immediately apparent what the drawing represents. Pictorial drawings are therefore used to represent objects in a more graphical and understandable way.

There are different methods of pictorial drawing. The two most popular are:

- isometric drawing
- oblique drawing.

Isometric drawing

Isometric drawing is used to display an object in 3D using three main base lines (sometimes called axes).

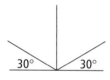

One vertical line and two receding lines at an angle of 30° are used to construct the object.

Therefore a simple object like a cube would look like the one shown on the left.

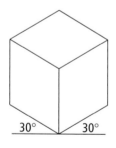

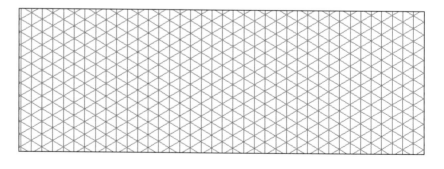

How to construct an isometric drawing

Isometric drawing can be quite difficult at first; the best way to learn is simply to have a go.

Using isometric paper (see above), redraw the isometric projection shown at the top of the next page.

Oblique drawing

The second principal method of pictorial drawing is oblique projection.

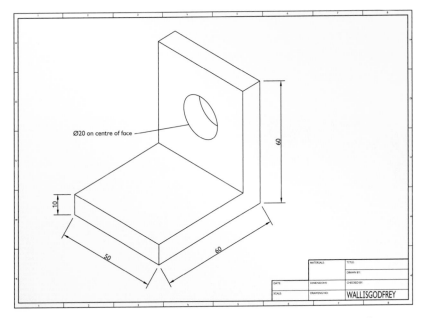

Oblique projection is simpler than isometric drawing as the front view is drawn parallel to the page – i.e. it is drawn flat, facing the viewer. Therefore features can be drawn in normally rather than at an angle of 30° – circles, for example, can be drawn in as circles rather than as ellipses:

- The first stage of oblique drawing is to produce horizontal and vertical lines to be used as the axes.
- A third receding line is then drawn in at 45°.
- The front view would be drawn as you see it to full scale.

Isometric hole (shown as ellipse)

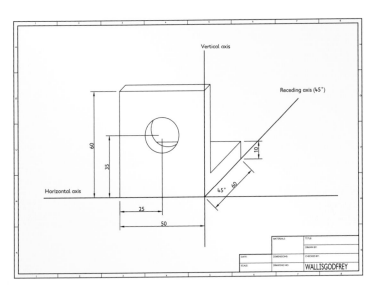

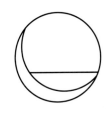

Oblique hole (shown as a circle)

- The sides, which are projected back at an angle of 45°, are drawn to half the real length. For example, if an object was 100 mm in depth it would be drawn to half scale, i.e. 50 mm deep.

Decreasing the length of the receding lines provides a more realistic view of the component. If the lines were drawn at full length, the receding features would look longer than they actually are.

Using the following tools construct the oblique drawing shown below:

- *pencil*
- *A3 or A4 paper*
- *rule*
- *45° set square*
- *drawing board and T-square (preferable).*

Tips:

Pick one of the sides to be your front view and draw this in as a 2D view.

Using your 45° set square, draw in the receding view at half scale.

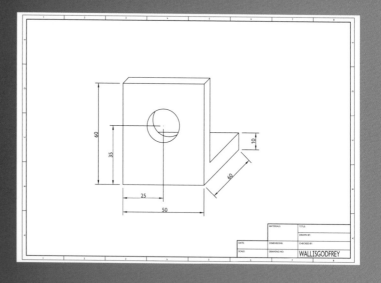

Calvin Klein, Ralph Lauren and Yves Saint Laurent are the names of some of the most famous designers in the world and are behind worldwide designs and fashions.

Graham Robertson and John Richmond are product designers in the design department at RPC Cresstale. They have the responsibility for making their ideas into real products that sell in their millions.

They take concept designs, sketches or design briefs from companies such as Avon, Body Shop, Virgin, Oriflame and many more cosmetics companies and bring them to life.

Concept designs are brought to life using Pro Engineer CAD software, and artwork and visuals with Photoshop, Freehand, Illustrator, and 3D Studio Max.

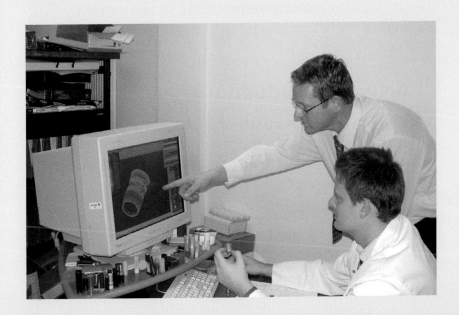

Graham Robertson and Neil Godfrey discuss the development of products using Pro-engineer

case study

RPC Cresstale

Graham Robertson has worked as a designer in many companies, including JCB, Caterpillar, Volvo and LMC Technik. His skills in the use of CAD, including Pro-Engineer, mean that he can design products from huge earth-moving equipment to the smallest cosmetics components.

John Richmond specialises in the graphic design of the products, giving the customers a chance to visualise the products long before they are made.

Although the products appear simple they are highly technical. Tolerances can be as low as ±0.01 mm. Pro-Engineer is a sophisticated CAD system that can produce perfect drawings with exact fits. Producing working drawings of the components is the first step.

Pro-Engineer (often referred to as Pro-E) gives fantastic visual images showing products in any colour and in any position. The system breaks down the drawing into digital information that can be used for many different applications required to produce the final products.

The team then sends digital information around Europe so that companies can build injection mould tools to produce the components. The digital information is also used to produce printing dies, polymers and also prototypes using a futuristic method of production known as stereo-lithography.

The electronic information is brought together in the design department to produce some of the world's leading cosmetics products.

An artwork template

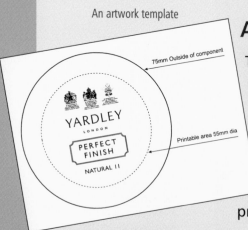

Artwork production

The artwork is designed and laid out in a drawing style which is sent electronically to the printing die manufacturers who will produce the printing dies for the product.

Inside the company, a dedicated team of toolmakers and technicians fine tune and commission injection-moulding tools, which will produce millions of products.

An injection-moulding tool can produce 12 components in one cycle, which takes approximately 12 seconds.

The production line uses injection-moulding machines to produce parts, robots for component transfer and specialist pneumatics systems to assemble the products. The designers become involved in the development of the production line itself, taking every opportunity to increase productivity and help the business stay competitive.

Injection-moulding tool

Pneumatic assembly machines assemble thousands of components per hour. This speeds up production and also reduces the chances of mistakes being made. Production workers are used instead in operations that are too complicated for machines.

Pneumatic assembly machine

The quality of the products is controlled by a quality team lead by the quality manager, Malcolm Wheatley, who is an expert in injection moulding quality assurance. A quality system known as BS 9000 ensures that the best possible system is in place.

To become designers for international fashion companies, earth-moving equipment, domestic goods, cars and high tech products for worldwide markets sounds almost unattainable, but it is not. Design engineers have trained so that they understand design briefs and product specifications. They have also trained in CAD, design and drawing techniques, subjects that you are probably already learning about!

Pneumatically driven devices pick and place lipstick covers in seconds

What's in this unit?

To complete this unit you will need to make an engineered product.

You will learn how to use a product specification so that you can understand clearly what is to be made and how you will make it.

You will learn to create a production plan based upon knowledge of many manufacturing processes, and you will learn how to distinguish clearly between mechanical, electrical, hydraulic and pneumatic systems and components. You will learn how to make informed choices to select appropriate components, parts and materials.

You will learn how to use hand tools and various industrial processes, including machining and welding, and be able to decide which processes will be the best ones to use to make your engineered products. You will learn to use a wide range of industrial tools, and how to work with them safely. You will also learn how to apply quality control techniques to industrial standards and to follow health and safety procedures.

Engineered Products

In this unit you will learn about:

Using a product specification

Introduction

If you have ever tried to make something such as a model aeroplane, bake a cake or make something you have designed yourself, you will know how difficult it is to make sure that everything goes in the right place at the right time.

Products made by companies can be single, one-off products or mass-produced products. This means that making sure everything is in place at the right time is even more difficult. Engineers need to **plan** how all the complex features of a product will come together to ensure that the product is made to the **product specification**, and therefore meets the demands of the customers who will eventually buy the product.

THE JARGON DRAGON

product specification – a list of all the important features of a product

Think
IT THROUGH

How would you react if you bought some trainers from a catalogue, and when they arrived they were not exactly how they were described? They could be a different colour or size, or even the wrong style.

The product specification

Product specifications can vary depending upon the product or the company that produces the product. The aim of a product specification is to list all the important features of a product so that the customer and the manufacturer know exactly what is to be made. If there are any disputes about the final product then the product can be checked against the description in the product specification.

A product specification should contain the following information:

- product description
- drawing references with measurements marked
- critical control points

- specification
- quality indication
- finish.

Example of a product specification – pneumatic air drill

Product		Pneumatic air drill A hand-held pneumatic drill powered by compressed air 80 mm long × 30 mm wide × 14.4 mm deep
Materials	Body	Aluminium
	Impellor	Aluminium
	Cover	Mild steel
	Twist drill	High-speed steel
	4BA screws	High-tensile steel
Measurement		All dimensions in mm Use working drawing 'Pneumatic Drill' 10-10-2003 Use assembly drawing 'Pneumatic Drill Assembly' 10-10-2003
Critical control points	Body	Check surface after casting Check shape after casting
	Impellor	Check diameter after CNC machining
	Cover	Check diameter after cutting
	Twist drill	Check diameter on receipt from stores
	4BA screws	Check length
	Assembly	Check that drill rotates in position by hand Check that drill runs by pneumatic force
Finish	Body	Main body sand-cast finish. Some pitting allowed. No short shots or flash
	Impellor	Machined surface finish
	Cover	Smooth finished with blackened finish
	Twist drill	Black finish as purchased
	4BA screws	Black finish as purchased
Quality indicators	Body	No sharp edges
	Impellor	No machine burrs. Smooth rotation in body bearing
	Cover	Even blackened finish. No sharp edges
	Twist drill	Straight shaft
	4BA screws	Each screw will screw into body smoothly
Safety	Body	Leather gloves, goggles, when sand-casting
	Impellor	Supervision on CNC machine centre
	Cover	Leather gloves, goggles, when blackening surface
	Twist drill	Be careful of sharp point when handling drill
	4BA screws	Ensure sharp edges are removed after sawing screws to size

Critical control points

As a product moves through the different stages of production it increases in value.

| £1.50 | £2.80 | £4.00 | £6.00 |
| Raw material | Cut wood | Assembly | Painted |

You should check that each operation has been undertaken correctly before moving on to the next operation, otherwise money and time are wasted making products that will not match the specification.

When the production process is extremely complicated, for example when making cars, it is impossible to check every single operation, but you can identify the most important points at which to check that the product is being made correctly. These are called **critical control points**.

Critical control points should be:

- when material is bought from the customer
- before an expensive operation
- after a difficult operation
- after an assembly operation
- when the product is finished.

THE JARGON DRAGON

critical control point – an important point during production when a component should be checked or inspected

quality – a measure of how well a product does the job that it is designed to do

Quality indicators

The **quality** of the products made by a company is extremely important. If the company produces poor quality products, then customers will complain or not buy them. The products could be dangerous, which could lead to further complications.

If products are made correctly to the product specification then the quality should be good. There are many definitions of 'quality' but you should always consider the following factors when deciding if a product is of good quality or not:

1   Can the product actually do the job it is intended to do?
2   Is the product made to the product specification?

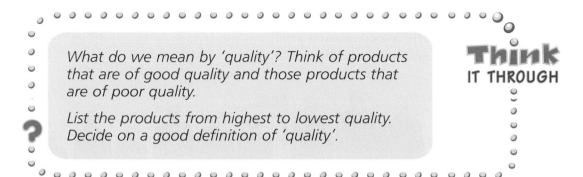

What do we mean by 'quality'? Think of products that are of good quality and those products that are of poor quality.

List the products from highest to lowest quality. Decide on a good definition of 'quality'.

Think
IT THROUGH

Measuring quality

When the quality of a product is measured, the kinds of measurements you need to make can be categorised in two ways. These are known as **variables** and **attributes**.

Variables

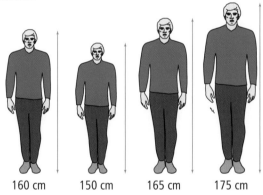

| 160 cm | 150 cm | 165 cm | 175 cm |

Variables are usually measured using numbers. The illustration above shows students' heights being measured. Each student has a specific height, and the students vary in height, so the measurements of height vary. Dimensions, temperature and speed are also examples of variable measurements.

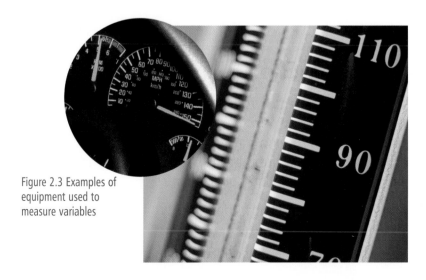

Figure 2.3 Examples of equipment used to measure variables

Attributes

When you measure attributes the answer is usually either 'yes' or 'no'; there is no variation.

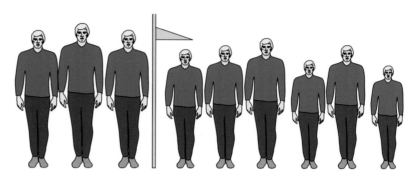

In the illustration above, the students will either fit under the height gauge or not. In the same way, a product could either fit under the gauge or not – yes or no, or pass or fail. This is a quick and easy method of checking the quality of components that are being mass produced.

Production planning

Planning the production of a product needs to be as detailed and precise as planning a military operation. Everything that needs to be done should be listed, and then engineers and planners need to work out how and when everything will be done.

The production plan

When you are given a product specification you will need to plan everything down to the last detail. This plan is know as the **production plan**, and it is the formal method used by engineers to make sure that the customers get what they are expecting. The production plan includes:

- materials, parts and components
- processes to be used
- tools, equipment and machinery

- sequence of production
- production scheduling
- how quality will be checked
- health and safety factors.

The important parts of a production plan are shown below.

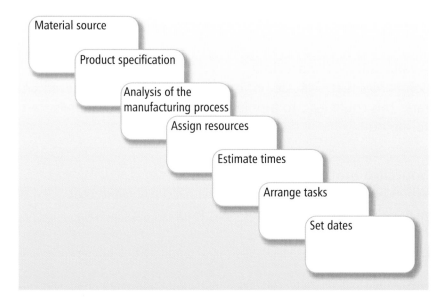

Material source

Product specification

Analysis of the manufacturing process

Assign resources

Estimate times

Arrange tasks

Set dates

Materials, parts and components
Materials

There are so many materials available that it is impossible to write about all of them. When selecting which material to use you should consider:

- the cost of the material
- the availability of the material
- the form in which the material will be bought
- what properties the material will need to have
- in which environments the product will be used
- the life of the product
- service requirements
- environmental effects
- skills needed to work with the material.

Cost of materials

Everyone wants their products to be made of the best quality materials, but this may mean that the price of a product is too high.

If you buy a lot of material from one company you will probably get a **discount**, so you need to consider how many products are being made.

Cost is always very important. Engineers look for materials that have the right properties and are good value for money. Sometimes buying cheaper materials to save money can cause problems with quality. It is easy to compare the costs of materials from various suppliers – simply look on their websites.

Availability of materials

Sometimes materials are not easily available; some may even be rare. This could lead to problems in production if materials start to run out.

Sourcing materials

When planning production, you will need to ensure that materials will be available for each stage of production. Materials are classified according to where they are used in the production process.

THE JARGON DRAGON

supplier – any business, shop, or person that provides materials to another business or customer

Classification of stock

- **Raw materials**
  These are the basic materials in the form in which they are bought in from a **supplier**. They are stored in the materials store and need to be protected from damage or theft.
- **Work in progress (WIP)**
  This means that the materials have been worked upon. They could have undergone one operation or many operations but are not yet the finished product.
- **Finished goods**
  These products are completely finished and inspected and are ready to be despatched, or sent, to the customer.

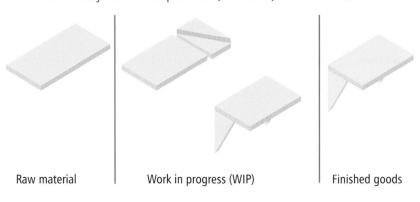

Classification of stock throughout the production of a shelf

| Raw material | Work in progress (WIP) | Finished goods |

Adding value to products

When material is worked upon, it costs money to do the work. This value is added to the cost of the product. Raw materials are the cheapest and finished goods are the most expensive.

Availability of stock

At each stage of production, the right type and amount of materials should be available to make the product.

Calculating material usage

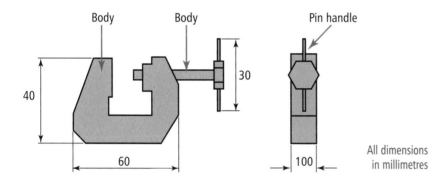

All dimensions in millimetres   Example drawing of a G-clamp

A simple G-clamp is manufactured with three components:

1 body
2 screw
3 pin handle.

One hundred G-clamps are to be made per hour cut from a bar that measures 40 mm wide × 10 mm thick × 2000 mm (2 metres) long.

- The body of the G-clamp measures 60 mm × 40 mm × 10 mm and is made from plain carbon steel.
- The screws are bought in boxes of 20 and have a 3 mm hole drilled in the head.
- The pin is 30 mm long, cut from lengths of a 3 mm diameter × 3 m long bar.

To ensure that there is enough material to make the G-clamps, you will need to make available the following amount of material per hour.

G-clamp material ordering

1 The body: 60 mm × 40 mm

$$100 \text{ components} \times 60 \text{ mm long} = 6000 \text{ mm or 6 m}$$
$$= 3 \text{ bars per hour.}$$

2   **The screws:** 100 components
      20 screws per box: 100/20 = 5 boxes per hour.
3   **Pin handles:** 30 mm long
      100 components × 30 mm = 3000 mm or 3 m
                             = 1 bar per hour.

If the screws are not ready, then the pins cannot be assembled. If the pins are not ready then the G-clamp cannot be finished. You need to recognise the importance of making sure materials and components are available at the right time.

THE JARGON DRAGON

process – the action of doing a task. It could be by an operator or a machine

Processes to be used

In a production plan you will need to consider the **processes** that will be used.

There are various types of processes, which include:

- assembly
- trimming
- joining processes
- finishing.

You will need to consider how suitable is the company's machinery for carrying out each process, and whether the workforce has the right kinds of skills. Engineers in industry may need to buy new machines or up-date existing ones, and they will need to consider the operators' skills and whether any special training is needed.

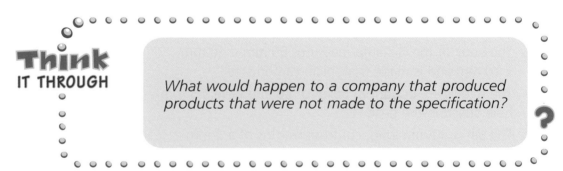

Think IT THROUGH

What would happen to a company that produced products that were not made to the specification?

Tools, equipment and machinery

Tools

When producing a production plan, you will need to consider all the tools that you have available. At this stage, it is very important that you understand the various groups of tools. You will look at individual tools later in this unit.

Marking-out tools

These tools often have sharp points that are able to mark steel. Some marking-out tools are steel versions of drawing equipment, so there are some straight-edged tools and others that are used to draw circles, angles or parallel lines. Only small forces are used with these tools.

Processing tools

These tools need teeth to remove material, or a head to bend metal. They need a handle that can be gripped firmly, as large forces are used to form the material.

Marking-out tools

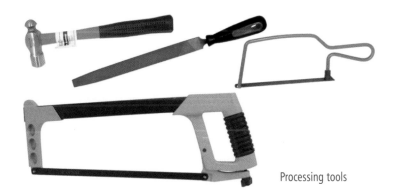

Processing tools

Measuring equipment

Most measuring tools have a scale on them that is used to read the measurement. Some use digital readouts, and some use dials. In tools where the measurement is made along an edge, such as a ruler, the edge that is used for measuring must be kept in good condition. Measuring tools often look like a clamp, so they can measure easily the distance across a component or part.

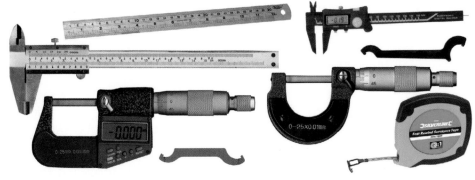

Measuring equipment

The tools that you select should be the best ones for the job. Using the wrong tools will cause problems with production and quality.

Which group of tools do each of these belong to?

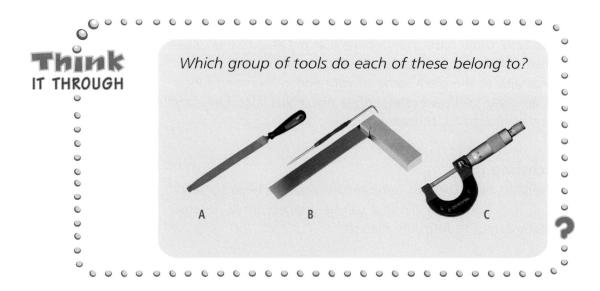

A　　　B　　　C

Solution

A This is a file. It has a firm handle and a set of teeth. It is a processing tool.

B There are two tools: a scriber, with a sharp point to mark fine lines on the material, and an engineer's square used to make lines at 90° to the side of a material's edge.
They are marking-out tools.

C This is a micrometer. It has two flat edges that fit over a component and a scale to read the measurement. It needs to be kept in good, clean condition.
It is a measuring tool.

Machinery

You will need to know about a range of machinery and be able to select which is the most appropriate for a particular job. You may need to take into account the cost of running the machine and the capabilities of the machine. You may even need to consider designing a new piece of machinery!

Sequence of production

Once you have decided upon which tools, machinery and processes are to be used, you can produce a **sequence of production**. Generally, all products are made following the same basic plan:

• prepare the materials
• process the materials

- assemble components
- finish the assembled product
- pack the product.

There must be a clear sequence of production to ensure that components are manufactured and assembled when required.

The sequence of production can be shown by various methods. Diagrams and/or short phrases are often used as these can display a lot of information in a clear and easy-to-understand way.

The sequence of production should contain details about the product and how it is to be made. Clear explanations of parts, processes and quality control techniques, including why they are being used, will help clarify the sequence of operation for the reader.

Sequence of production for an instrument vice

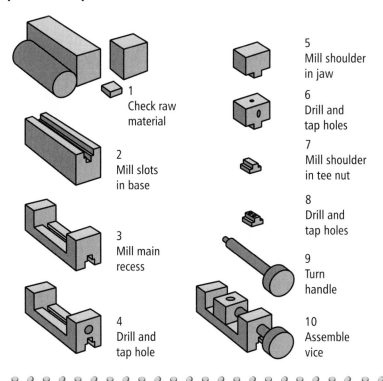

1 Check raw material

2 Mill slots in base

3 Mill main recess

4 Drill and tap hole

5 Mill shoulder in jaw

6 Drill and tap holes

7 Mill shoulder in tee nut

8 Drill and tap holes

9 Turn handle

10 Assemble vice

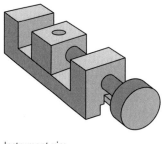

Instrument vice

Think IT THROUGH

Where are the critical control points for the production of the instrument vice?

Example of sequence of operation of instrument vice, using sentences

Prepare materials

Clean the material.
The material should be checked for size to ensure that it is big enough to make the product.

Processing

De-burr all material using a hand file.

Base
Mark out the base using a surface table and scribing block.
Mill the recess in the base using a vertical milling machine.
Slot drills are used which can cut long slots easily.
Mill jaw recess using a horizontal milling machine. This machine can remove large volumes of material.
Case harden the fixed jaw as material cannot be hardened by heat treatment alone.

Jaw
Mark out the jaw using a surface table and scribing block
Mill shoulder of jaw using a vertical milling machine.
Drill and tap top hole and drill hole for shaft
Surface grind front of jaw. This process is much more accurate than milling.

Case
Harden jaw face.

Tee-nut
Mark out the tee nut using a surface table and scribing block.
Mill shoulder using vertical milling machine.
Drill and tap holes. Use hand vice, as the torque forces are low.

Handle
Turn main shaft of handle on centre lathe. Use a right-hand cutting tool.
Use a die and die holder supported by the tailstock of the centre lathe to form thread of the shaft. This will keep thread true and straight.
Use a knurling tool to form the texture of the handle.

Assembly

Place jaw in the slot of the base. Ensure that the jaw can slide.
Attach the tee nut to the jaw and ensure that the jaw can still slide.
Attach handle to jaw and secure with grub screw.
Close jaw and then open fully to ensure movement.

Finish the assembled product

All components of the instrument vice should be coated in a thin film of oil.
This will prevent rusting.

Packing

Place in cardboard container with label.

Here we see the sequence of operation for an angle plate and bracket.

Using a diagram with notes method

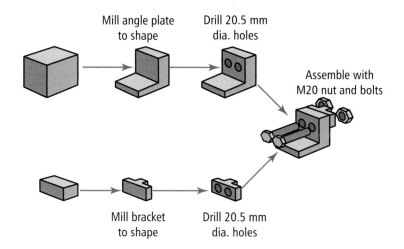

Angle plate and bracket sequence of operations

Using a block diagram method

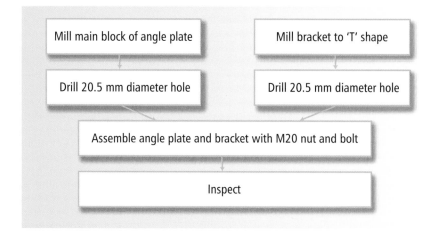

Planning sheet

A **planning sheet** is the formal way to write down everything that relates to the manufacture of a product. For each component, details of every operation are described. These include information about the process, materials, tools, equipment, health and safety, quality and the time it takes to complete the operation.

Each component has its own planning sheet, which helps ensure that the whole product will be completed correctly and safely.

The planning sheet should include the name of the product and the name of the person who wrote it.

Example planning sheet

Student name   Date   Tutor

Name of part: Pneumatic drill body

Stage	Process	Materials	Tools and equipment	Machinery	Health and safety	Quality	Time checks
1	Prepare mould		Wooden pattern Casting box Casting sand Trowel		Goggles Watch for sand on floor	Check cavity is correct	30 min
2	Melt aluminium (660°C)	Aluminium (bar and off-cuts)	Ladle	Hearth	Teacher or technician to do this	Aluminium should run easily	15 min
3	Pour molten aluminium		Ladle		Wear leather apron		5 min
4	Cool and remove		Tongs Leather gloves		Aluminium may be hot	Check for short shots pitting	10 min
5	Mill top surface flat	Casting	20 mm diameter Slot drill	Vertical milling machine	Wear goggles	Body is 14.5 mm deep	30 min
6	Mill pocket and drill holes	Casting	10 mm diameter slot drill C/Drill 3.3 mm diameter drill	CNC machine centre	Wear goggles	Diameter of pocket 15 mm	15 min
7	Drill input and exhaust holes	Casting	3.9 mm diameter drill	Pedestal drill vice	Wear goggles	Check location of holes	15 min
8	Tap 4 BA holes	Casting	4 BA taps Tap wrench Bench vice		Test thread with 4BA screw		15 min

Gantt charts

A **Gantt chart** is a type of diagram used to help plan activities or projects. It is a good way to help you plan how long a product will take to manufacture.

This technique will allow you to give realistic deadlines that will take into account the processing and assembly times. In industry, once engineers are committed to a deadline, they have to stick to it. There could be serious consequences if they miss it, including:

- losing future orders
- not being paid for the late work
- losing their reputation for delivery
- they may have to pay a large penalty to the customer who was expecting the goods by a certain date.

Have you ever been let down by someone who said they would do something by a certain date but failed to do so? How did it make you feel?

Think
IT THROUGH

Example plan

If you were planning to make a cup of tea, you would put the kettle on but you would not have to wait for it to boil before putting the tea bags in the teapot – this could be done while the water was heating up.

If you prepared the teapot and put milk and sugar in the cup, and then put the kettle on, you would waste time. When planning, you should always try to make best use of the time available. A Gantt chart allows you to do this easily.

Operations:

- fill kettle
- boil kettle
- put teabag in teapot
- put milk and sugar in cup
- pour boiling water in teapot
- pour tea.

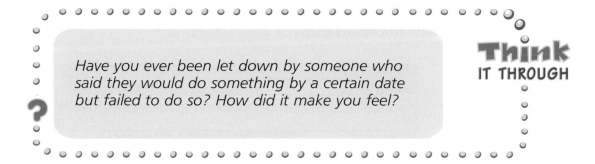

Stage	Operation	1	2	3	4	5	6	7	8
1	Fill kettle								
2	Boil kettle								
3	Put teabag in teapot								
4	Put milk and sugar in cup								
5	Pour boiling water in teapot								
6	Pour tea								

Gantt chart to produce a cup of tea

The column on the left lists the operations, and the top line shows the time (in minutes). The blue blocks show when work is being done.

Engineers and planners find this technique extremely useful. It can be used in all sorts of situations, for example planning a new stadium or a tunnel, or even a trip to Mars! There could be hundreds of operations involved, and the time could be spread over many months or even years. With very complicated projects like this, an IT-based system is usually used.

Scheduling using a Gantt chart is very straightforward when using **Microsoft Excel** or other spreadsheets.

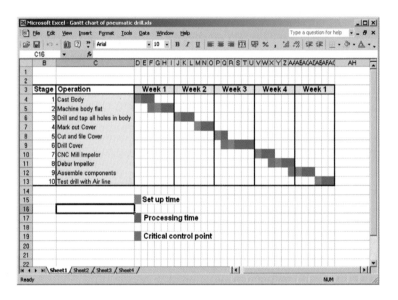

There are also specialist programs such as **Microsoft Project**, which are specially designed to help businesses with project management and accounting.

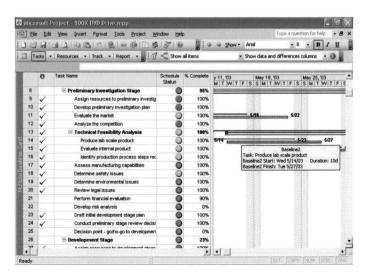

Production scheduling

When the plan is in place, you need to consider a time scale for the manufacture of the components for the product.

You will need to consider:

- how long each operation will take
- which operations need to be done before other operations can take place
- whether any operations can be done at the same time using different operators or machines
- the deadline for completion of the job
- whether you have to rely on others to complete the work (these could be suppliers or sub-contractors)

Example Gantt chart (based on times from pneumatic air drill planning sheet)

Stage	Operation	Week 1	Week 2	Week 3	Week 4	Week 5
1	Cast body					
2	Machine body flat					
3	Drill and tap all holes in body					
4	Mark out cover					
5	Cut and file cover					
6	Drill cover					
7	CNC mill impellor					
8	Debur impellor					
9	Assemble components					
10	Test drill air line					

Each operation is broken down into times. In this example each week is broken down into 15-minute blocks to help planning. Set up times here are shown in red, and processing, or making, times are shown in blue. Green squares show critical control points.

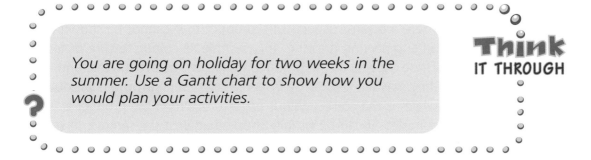

? *You are going on holiday for two weeks in the summer. Use a Gantt chart to show how you would plan your activities.*

Think
IT THROUGH

How quality will be checked and inspected

As the products are made, they will need to be checked. This process is known as **inspection**.

You will need to consider:

- what you are checking against
- which aspects of the product will be inspected
- how often the components will be inspected
- who will be responsible for inspecting the components
- what happens to bought-in goods and materials.

Inspecting against the drawing

The engineering drawing gives the sizes of all the components. Before production begins, the manufacturer and the customer agree that the drawing is correct. The drawing is given a unique number, which is located on the drawing. This ensures that when the product is inspected it is compared with the correct drawing.

If the drawing is altered or modified at any stage, then the drawing is given a modification number. The first modification may be called 'Modification A', the second may be 'Modification B', etc. It is important to destroy all drawings that are not the current modification.

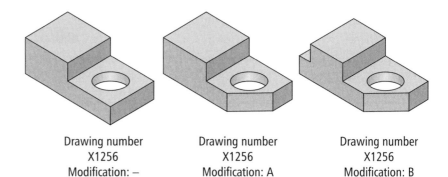

When a component is modified it is given a new modification number

| Drawing number X1256 Modification: – | Drawing number X1256 Modification: A | Drawing number X1256 Modification: B |

Inspecting against the specifications

There are some aspects of a product that cannot be shown on an engineering drawing, for example its colour, texture or taste. These aspects of the product are known as **attributes**, and they may all need to be inspected. Often a **sample** is produced which is the correct colour or texture. When the product is complete it can be visually checked against the sample. In some cases special measuring devices are used that measure colour and texture scientifically.

It is almost impossible to manufacture colours to an exact shade, so usually there are two sample colours, a lighter shade and a darker shade, giving a range of shades that is acceptable.

THE JARGON DRAGON

sample – a component or product that has been specially produced or selected to represent the acceptable quality for manufacture

Which of these products are within tolerance? These products must be darker than the light sample and lighter than the dark sample.

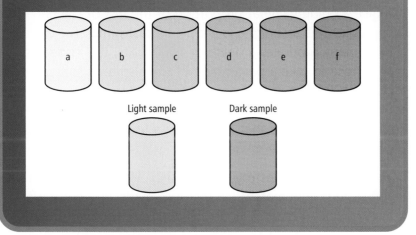

Light sample   Dark sample

As you can see, it is much more difficult to measure attributes!

Discuss your findings in a group. Did you find it difficult?

Think
IT THROUGH

Deciding which aspects of the product will be checked

It is impossible to check every single dimension of every component. Inspection takes time and is expensive. Planners need to find a balance between ensuring that the product is inspected sufficiently, and that the inspection process does not take up too much time or expense.

In industry, planners need to work with the **quality manager** to ensure that when something needs to be inspected, the inspectors have the instruments, skills and time to measure the components accurately.

Many products have a specification that relates to their design and performance requirements, which is laid down in British, European or International Standards. If a product complies with the appropriate published specification, the manufacturer can apply for British Standard Kitemark approval. Products carrying the Kitemark can generally be relied on to be of good quality.

THE JARGON DRAGON

quality manager – a person responsible for the overall quality of the products produced by a business

How often will products be inspected?

If every single product is tested this is known as 100% inspection. It is time consuming and laborious and can lead to a lack of concentration by the inspector. In most cases, therefore, a small sample of the products may be tested. This will give a guide to what proportion of the rest of the batch is acceptable. This technique is known as **sampling**. For example, if a company produces a batch of 1000 bolts, the inspector may choose to take a percentage of them to test. This could be 10%, or 100 bolts. If they are picked at random and are all satisfactory, then the whole batch will be seen as satisfactory. This, however, means that if there are a few products that are incorrect they may be missed.

Some products are destroyed when they are used. For example, if you tested every matchstick that was made to see if it works, there would be nothing left to sell. On the other hand, there are some products where 100% inspection is essential, such as in the aviation industry where safety is a major issue. There are often lots of difficult decisions to be made concerning how often, and how many, products should be tested.

ACTIVITY

Think of two different products.

- *How many products per hundred should be inspected?*
- *Do they need to be destroyed on inspection?*
- *Could you use a 'Go' and 'No-Go' system?*
- *Which of these instruments could be used to inspect the products?*
- *(Use the comparison chart on the next page to help choose the measuring equipment.)*

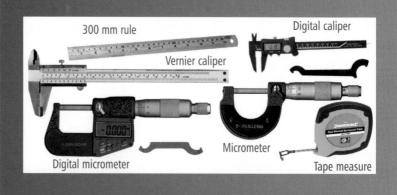

300 mm rule

Digital caliper

Vernier caliper

Micrometer

Digital micrometer

Tape measure

Planners need to be aware of the range of measuring equipment available, and what each instrument can measure. There are a number of features to take into account when choosing the most suitable measuring instrument to use:

- **Accuracy:** this is how close the measured size of a component is to its actual size.
- **Repeatability:** this is the ability of an instrument to give the same reading every time it is used to measure the same thing.
- **Range:** this is the difference between the smallest size an instrument can measure and the biggest size it can measure.

A sample of measuring equipment with measuring characteristics

	Accuracy	Repeatability	Range	Cost
Spring calipers	0.5 mm	Good	0–120 mm	Cheap
Rule	0.5 mm	Good	300 mm	Cheap
Vernier height gauge	0.05 mm	Good	250 mm	Very expensive
Vernier caliper	0.05 mm	Good	150 mm	Expensive
Digital caliper	0.01 mm	Good	150 mm	Very expensive
Micrometer	0.01 mm	Excellent	25 mm	Expensive

Drawings give **dimensions** and **tolerances**, which help determine which measuring equipment to use.

Health and safety factors

All engineering and manufacturing companies need to follow the Health and Safety at Work Act 1974 (HASAW). This relates to everyone in the company, whether they are employed or not. When production plans are being produced, the planner must ensure that they have made the production process as safe as possible.

They should consider:

- all personnel
- storage
- machinery
- tooling
- all transport
- materials handling equipment
- hazardous substances.

Safety issues could include having a safe working area, guarding machines or training employees to undertake a particular task.

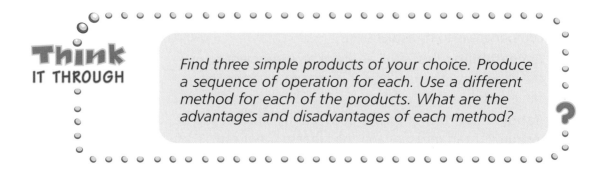

Think IT THROUGH

Find three simple products of your choice. Produce a sequence of operation for each. Use a different method for each of the products. What are the advantages and disadvantages of each method?

Choosing components, parts and materials

Materials and their properties

Engineers have an almost endless range of materials to choose from. These materials fall into different categories. It is important to understand these different categories so that you can quickly identify the best material for the job.

Before you choose a material you need to know some important information about the product that is to be made.

- **Service:** What does the product need to be able to do?
- **Properties:** What characteristics will the material need to have, such as strength and weight?
- **Environment:** What sort of surroundings will the product be used in?
- **Aesthetics:** What will the product look like? Is it important that it has a particular look, colour or feel?

Some products can be made from different materials such as car body parts, cups, furniture and tools. Therefore there is not always one perfect material for the job.

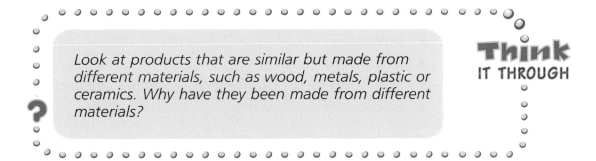

Look at products that are similar but made from different materials, such as wood, metals, plastic or ceramics. Why have they been made from different materials?

Think IT THROUGH

Putting materials into categories

Engineering materials fall into four main categories:

- metals
- polymers
- ceramics
- composites.

Metals

Metals fall into two main categories, which are **ferrous** and **non-ferrous** metals.

Ferrous metals contain iron. Pure iron has some uses but is generally too soft and **ductile** for most engineering products. However, if a small amount of carbon is mixed with the iron this changes the material's properties. Adding more carbon to steel allows the steel to change its properties when it is heat-treated.

Adding carbon to iron makes a new material known as **carbon steel**. Carbon steels can be categorised depending on how much carbon is in the steel. The categories of carbon steel are:

- 0.1%–0.3% carbon: mild steel
- 0.3%–0.7% carbon: medium carbon steel
- 0.7%–1.3% carbon: high carbon steel.

Grey cast iron has 94% iron and 3% carbon and small amounts of other elements. It is a hard material but melts relatively easily so it can be melted and poured into moulds to make products. Cast iron can be machined quite easily after it has been moulded. A disadvantage of cast iron is that it can be brittle. This means that if a large force is applied it could snap or shear.

Non-ferrous metal do not contain any iron. There are many non-ferrous metals, some of which are very common. You may already recognise some names such as copper, brass or aluminium.

THE JARGON DRAGON

carbon steel – a mixture (or alloy) of iron and a small amount of carbon. It is commonly used in engineering

Non-ferrous metals include:

- aluminium
- copper
- many alloys (brass, bronze, phosphor-bronze, gun metal)
- precious metals (silver, gold).

Aluminium

Aluminium is a light silver colour. It is lightweight but is relatively strong. It has a melting point of 660°C and can be moulded easily into shape. This material does not corrode easily. Aluminium can be made into thin sheet or foil.

Aluminium is used for:

- aeroplane components
- mountain bike components
- foil for cans
- ornaments.

Copper

Copper is dark orange in colour and is very commonly used throughout the world. It is very ductile, which means that it can be stretched into very thin lines to make copper wire. It is excellent for conducting electricity and heat, and can be easily worked upon in the workshop, or pressed into shape by machines. It does not corrode easily, but over long periods of time the surface will oxidise. This oxidation can be seen as a light green, dusty covering. Copper melts at 1080°C.

Copper is used for:

- copper wire
- copper water pipes
- plumbing components such as pipes and joints
- coins
- saucepans
- scientific instruments
- ornaments.

THE JARGON DRAGON

alloy – a mixture of two or more metals or elements

Alloys

When metals or elements are mixed together to form a new material the new mixture is known as an **alloy**. Alloys are classified as **light alloys** or **heavy alloys**. Light alloys include zinc, aluminium and titanium alloys. Heavy alloys include copper, lead and nickel alloys. Modern engineering alloys contain many elements and the process of designing new materials is extremely complicated.

Brass

Brass is an alloy of copper and zinc. It has a low melting point (800°C) and can be cast easily. Brass does not corrode easily and can be polished to give a high gloss finish similar to gold. This means that it is often used for ornamental or decorative furniture for indoor and outdoor use. Brass can be used for:

- decorative fittings
- musical instruments
- electrical fittings
- locks and keys.

Bronze

Bronze is an alloy of copper and tin. There are two main types:

- Gun metal is 88% copper, 10% tin and 2% zinc. It is relatively easy to machine and is corrosion resistant. It is used for marine components such as valves, pumps and steam fittings.
- Phosphor-bronzes can vary according to the quantities of each alloying metal. They also contain the chemical phosphorous. Phosphor-bronzes have excellent wear resistance and corrosion resistance and are used to produce bearings.

Precious metals

Metals such as gold, silver and platinum are very precious and so are very expensive. Although these metals are very often used to make jewellery, they also have other uses. For instance, silver and gold are extremely good at conducting electricity.

Alloys have been used for thousands of years

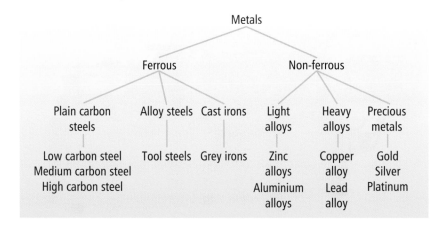

An overview of some of the types of metals available

Think
IT THROUGH

Consider the following components. Try to select an appropriate metal for each. Remember, there could be more than one answer, so discuss your answers in small groups.

- *an outside door handle*
- *mountain bike pedals*
- *water pipes*
- *a workshop vice*

Polymers

'Polymers' is the scientific term used to describe plastics.

The first plastics were developed over 100 years ago, and although they had limited uses, by the time of World War 2 many new applications had been found, such as in clothing.

Since that time scientists and engineers have found countless uses for polymers. Today, thousands of products are being made using polymers.

Polymers are light, corrosive-resistant and can be coloured. Products can be manufactured easily and quickly, once tooling has been produced. Tooling or mould tools are needed to produce components from polymers. This tooling can be very expensive.

An **injection moulding** machine is used to form plastic injection-moulded products. These machines are used in mass production and can sometimes take only a few seconds to produce complex components such as mobile phone covers.

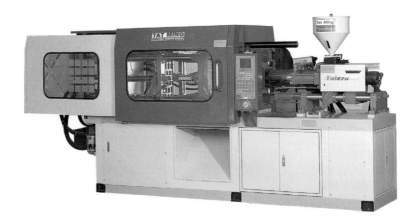

There are many different types of polymers and these can be categorised into two main groups: **thermoplastics** and **thermosets**.

Thermoplastics

When heated, thermoplastics can take on a new shape if pressure is applied. If the product is heated once more, the material can be re-shaped using pressure.

Thermosets

Thermosets are also formed into shape using heat. When the shape is cooled, however, the component that has been produced cannot be re-shaped in the same way as thermoplastics. Thermoset polymers can withstand higher temperatures than thermoplastic polymers, and are used for products where there is a danger from heat. These products include many electrical components, such as switches and light fittings, or handles for products that give off heat, such as kettles, pans, irons, etc.

Phenolic

Phenolic used to be often used for radio and TV cases as it has good heat and electrical resistance. It is also very rigid. It is now used less frequently, but can still be seen in some electrical installation products.

Polyester resins

Polyester resins are more commonly known as glass-fibre and are used for car bodies, boats and canoes. They are extremely strong and are suitable for most applications where strength is required. They are often used to make large products or products with large surface areas, as they are not moulded using injection moulding.

Epoxy resins

These are used to make adhesives and also glass-fibre laminates.

Natural polymers

Rubber is a natural polymer that can deform quite significantly. It can be stretched or compressed and when the load is released the rubber returns to its original shape. This property is known as **elasticity**.

Polymer additives

Although there are many types of basic polymer, additives can be introduced to change their properties. This gives an almost endless range of polymers that can be used by engineers and designers.

Additives include:

- plasticisers
- pigments
- fillers.

Plasticisers

Plasticisers help soften polymers, for example PVC (polyvinyl chloride).

Products such as drain pipes and guttering can be made from PVC, as can clothing, such as imitation leather.

Pigments

Pigments give a product its colour. Products can be made in exactly the same way, and different pigments used to give a variety of colours. Examples would be pens or pieces in a board game.

Fillers

Various fillers can be added to components. The fillers change some of the component's properties, such as strength.

Choosing polymers

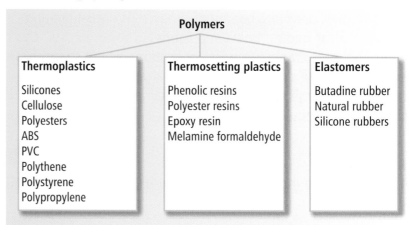

Thermoplastics	Thermosetting plastics	Elastomers
Silicones	Phenolic resins	Butadine rubber
Cellulose	Polyester resins	Natural rubber
Polyesters	Epoxy resin	Silicone rubbers
ABS	Melamine formaldehyde	
PVC		
Polythene		
Polystyrene		
Polypropylene		

Thermoplastic
ABS (acrylonitrile butadiene styrene)

ABS material has excellent resistance to impact. It is used to manufacture telephones, helmets and surfboards. If a product needs to be made accurately but will be expected to take impacts in normal use, ABS is a suitable material with which to make it.

PVC (polyvinyl chloride)

PVC is a very versatile material. In solid form it is strong and durable. It is used to make windows and conservatories as it can be extruded easily. With fillers it can be made soft and is often used as an alternative to leather.

PVC has excellent electrical resistance properties and is used as an insulator to cover electric cable.

Polystyrene

Polystyrene is hard and strong. It can be brittle in thin sections but has the advantage of being able to be manufactured in transparent or clear form. It is used as a safe substitute for glass such as in clear covers, boxes or pens.

Nylon

Nylon is very tough and has good wear resistance. It can be moulded very accurately and is used for making technical products such as gears and bearings.

Perspex

Perspex is supplied in flat sheet and often used as a safe alternative to glass. It is used for machine guards and boxes. It is often transparent but can be coloured for a better visual effect.

Mechanical components

When components are to be assembled there are many methods of keeping them attached. There is a wide variety of mechanical devices know as **fasteners**.

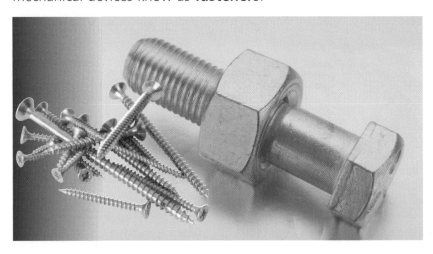

Nuts, bolts, rivets, screws and washers are all examples of fasteners, but there are so many different types available that it is impossible to list them all. Engineers have to consider all of the fasteners available and choose the best ones for the job. When deciding which fasteners to use you need to consider the following features:

- threaded fasteners such as bolts require nuts, or the component that they go into needs to be threaded
- unthreaded fasteners are known as rivets and are squashed over at the end to form the joint.

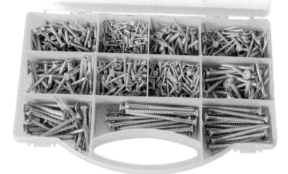

Examples of threaded fasteners (left) and unthreaded fasteners (right)

A bolt will need a nut, which is an extra component to buy, store and account for. However, they are very strong and are reversible, which means the component can be taken apart easily. Sometimes no nut is used and the component has a screw thread formed into it. This means that, instead of making a simple hole in the component, an operation is needed to thread the hole.

The important features of a bolt

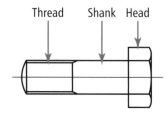

Thread   Shank   Head

Head style

There are many different head styles. Some are domed for smoothness, others are flat so they are flush with the surface. Here are some examples.

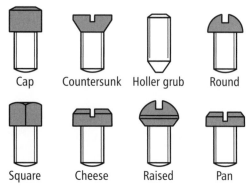

Cap   Countersunk   Holler grub   Round

Square   Cheese   Raised   Pan

Sometimes the shape of the head is not important, so there may be more than one type that will do the job. In such a case the choice of fastener can be based on cost or availability.

Head type

The fastener will need to be turned with some type of tool, such as a screwdriver, wrench or hexagon key. This means that the head has to have the correct slot, shape or recess.

Hexagon   Slotted   Pozidriv   Phillips   Socket   Splined

There are many different head types. It is useful to ask a supplier of fasteners or look in a catalogue to see the head types available. Most suppliers have excellent websites which give details of their products.

> *Try to design a head type of your own, but remember also to design the tool that is to be used with it.*

Think
IT THROUGH

Size

The length or diameter of a bolt can affect the overall product or assembly. Engineers need to choose the best size that will give the necessary strength, but not add weight or expense.

Here we see three plates with holes to be bolted.

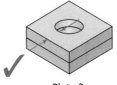
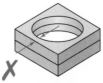

Plate 1
The bolt may be weak

Plate 2

Plate 3
The plate is weak

- **Plate 1:** The bolts may be too small and will shear or snap.
- **Plate 2:** The hole leaves material on each side so that the plate or bolt does not shear.
- **Plate 3:** The bolt will be strong, but there is little material left, which could mean that the plate breaks easily.

Thread type

There are many different thread types.

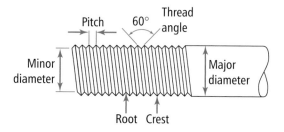

The most common screw thread form is the one with a symmetrical V-profile. Threads can also be square, acme or buttress. The included angle of V-profiled threads is 60°. This form is used in the ISO/metric thread.

Pitch

The pitch is the distance between the high points on a thread. This can vary. The pitches of threads must be the same for threads to go together. The thread type gives the thread its shape and size. Thread types are usually standard sizes.

Metric threads come in a range of sizes and diameters – these are 2 mm, 4 mm, 6 mm, 8 mm, 10 mm, etc., and are coded with an upper case 'M' before the diameter. For example, a 6 mm diameter metric thread is written as M6. If the thread had a pitch of 1.00 mm it would be written M6 × 1.00.

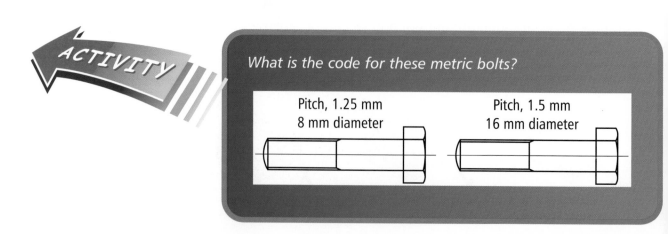

What is the code for these metric bolts?

Pitch, 1.25 mm
8 mm diameter

Pitch, 1.5 mm
16 mm diameter

ACTIVITY

Coating (or plating) of threads

The external threads should not be greater than basic size after plating, and the internal threads should not be less than basic size after plating.

Engineers should also consider the material and the finish, such as a galvanised finish.

It can be seen that choosing the best bolt can be quite a challenge.

Remember when choosing threads to consider:

- head type
- head style
- thread type
- length
- coating
- diameter.

Washers

A washer is a disc of metal with a clearance hole larger than the bolt. Washers are used for a number of reasons.

Washers can prevent the nut and bolt from unloosening due to vibration. They spread the force created over a larger area, in other words they spread the load. They can also prevent damage to the surfaces being tightened together, as the bolt head or **nut** will turn against the metal washer and not the material surface.

THE JARGON DRAGON

nuts – these can be hexagonal, square or any number of shapes with an internal thread for use with bolts

There are many kinds of fasteners. Use the Internet to identify suppliers, and look at the range of products available.

Think
IT THROUGH

Rivets

Rivets are unthreaded fasteners which are similar to bolts but have no thread.

Here we see the process of how two plates are joined with a rivet:

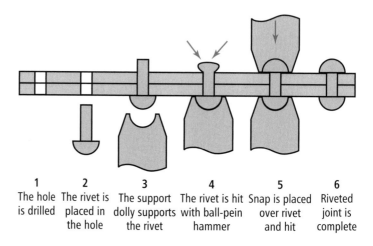

1	2	3	4	5	6
The hole is drilled	The rivet is placed in the hole	The support dolly supports the rivet	The rivet is hit with ball-pein hammer	Snap is placed over rivet and hit	Riveted joint is complete

Round head   Coutersunk head

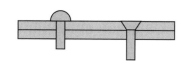

Rivets have head types similar to bolts. Here we see two common types.

The advantage of the countersunk head is that it lays flush with the work surface. The countersunk rivets can be filed to appear invisible. The disadvantage of the countersunk head is that the hole needs to be countersunk, which is an extra operation.

Hydraulics and pneumatics

Hydraulic and **pneumatic systems** use fluids to transfer forces from one place to another. BS 2917 is the standard in use when drawing the symbols.

The fluid can be gas or liquid. In systems that use gas, air is used. These systems are called pneumatic systems. In systems where incompressible liquid is used, the system is known as a hydraulic system.

Gases can compress. In a pneumatic system the air is compressed and the air pressure this creates is used to do the work. Liquids do not compress, and are used in systems where much heavier work is required.

Example of a pneumatic system

If we blow up a balloon, we could release the air into a system which would move a piston (see the diagram on the next page).

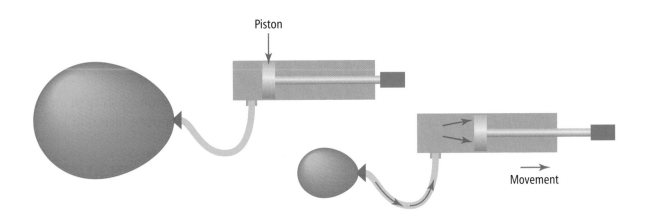

Piston

Movement

Hydraulics

If you use the system shown below, which is filled with liquid, and pushed down on the small piston (effort), the load would rise. At first sight, the diagram gives the impression that something small has lifted something big with no loss. However the small piston has had to move twice as far. What has been gained in force must be sacrificed in distance travelled by the piston.

The tube, where the effort is supplied, is thinner than the load tube. This means that the effort will move twice as far as the load, but the system could lift twice as much load as the effort supplied.

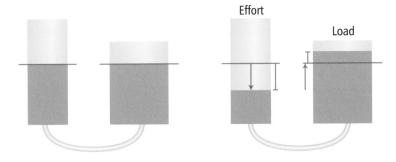

Effort

Load

Movement of fluid in a hydraulic system

Pneumatics
Actuators

Actuators convert the air in a pneumatic system into physical movement. This is either rotary (spinning around) or linear (moving in a line).

Pneumatic cylinders

A cylinder is a linear-type actuator. Cylinders have numerous functions and are often used in machinery to stamp or move components.

The air pushes a piston. In the **single-acting cylinder** the piston has to be moved back with some other force such as a spring. When the air pressure is removed the spring moves the piston back.

A **double-acting cylinder** pushes the piston open with air and pushes the piston back with air in the opposite direction.

Pneumatic systems use a series of valves which allow the air to switch direction. In this case, the piston can be pushed back.

A pneumatic cylinder

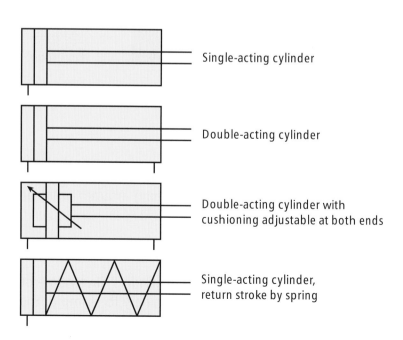

Symbols used to represent types of cylinders

Valves

In pneumatics, the compressed air flows along an air line. If the direction of the air needs to change, then a valve is drawn showing two or more blocks with arrows inside. Each block shows the direction in which the air will flow.

Ports are the connections where air enters or leaves the valve.

In Figure A, compressed air enters the valve through the **air inlet** port but comes to a stop as it has nowhere to go.

When the push button is pressed (Figure B) the valve switches so that the air now flows through the valve.

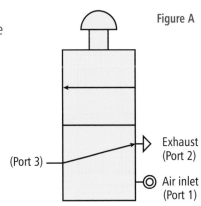

Figure A

(Port 3)

Exhaust
(Port 2)

Air inlet
(Port 1)

In this diagram, the valve is shown in both positions. As there are two positions that the valve can be in, this is known as a **two-way valve**.

We can see what would happen to the air if we imagine the block moving to the open position.

Position A
Open normal position

Position B
Closed normal position

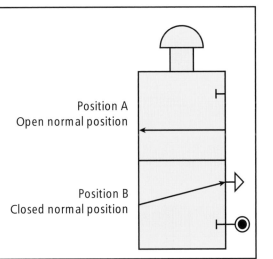

Figure B

Button is pressed

Exhaust

Air inlet

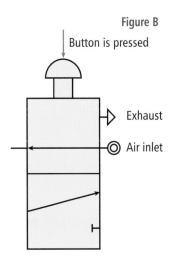

Think
IT THROUGH

Consider the two examples shown below.

For each example, say where the air will be directed to when the directional control valve is activated.

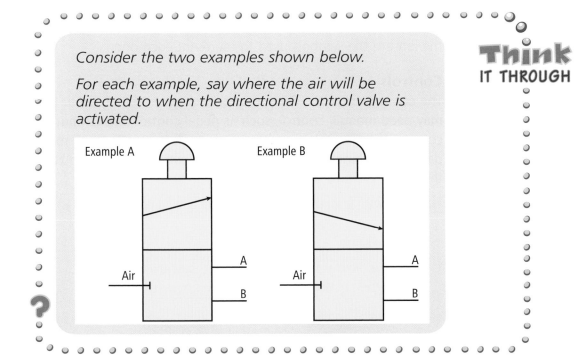

Example A

Example B

Air

A

B

Air

A

B

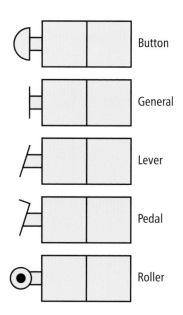

Button

General

Lever

Pedal

Roller

Valve symbols – classified by method of introducing air

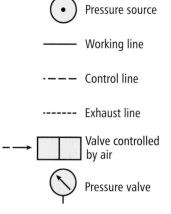

⊙ Pressure source

—— Working line

·–·– Control line

······· Exhaust line

Valve controlled by air

Pressure valve

Other important pneumatic symbols

Solutions

Example A: The air will go to port A.
Example B: The air will go to port B.

Valve classification

Valves are classified by how many ports they have and how many position they can be in.

The format is:

Number of ports / Number of positions (shown by blocks).

An example could be a 2/2-way valve.

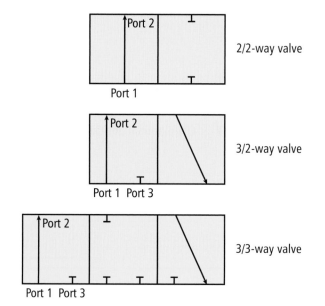

2/2-way valve

3/2-way valve

3/3-way valve

Air can be introduced to a valve by a number of methods. On the left are the symbols used to represent each one.

Controls

The controls used will depend upon the system being used. It may need manual control, such as pedal control or push button control. These may be used to clamp tools, open doors or move parts of machinery. Pneumatic systems save manpower by using the air pressure to supply power.

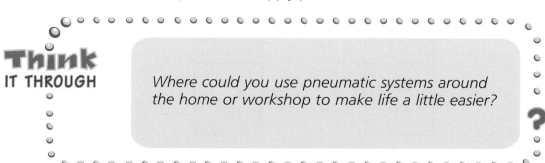

Think
IT THROUGH

Where could you use pneumatic systems around the home or workshop to make life a little easier?

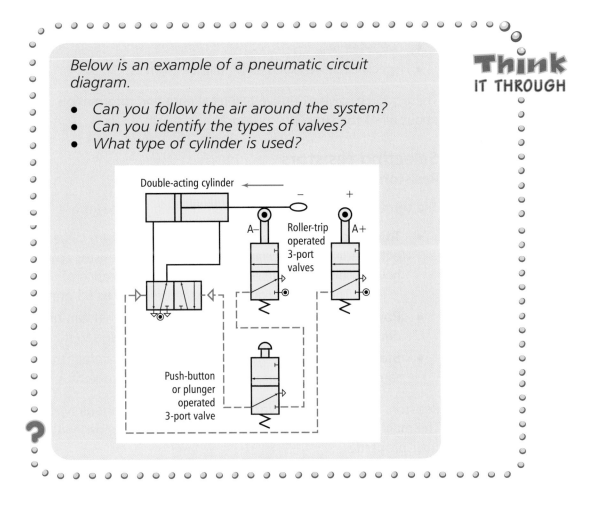

Below is an example of a pneumatic circuit diagram.

- *Can you follow the air around the system?*
- *Can you identify the types of valves?*
- *What type of cylinder is used?*

Double-acting cylinder

A−   Roller-trip operated 3-port valves   A+

Push-button or plunger operated 3-port valve

Think
IT THROUGH

Electrical and electronic components

In order to produce an electronic circuit you will need to understand the various types of electrical equipment and their symbols:

- resistors
- thermistors
- potentiometers
- capacitors
- inductors
- diodes

- LEDs
- cables
- batteries
- motors
- audible alarms
- lamps.

Unit 1 gives a description of each of these electrical components. This unit deals with how to select the best components.

Designer and engineers need to choose the best components for products. There are many sources of this information, including:

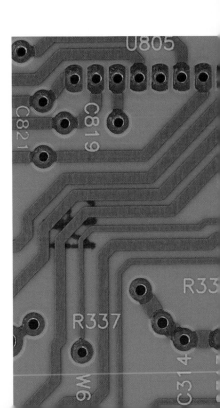

- catalogues
- websites
- specialist stores such as Maplin Electronics.

These sources give extensive technical data and information about all of the suppliers' products.

Selecting resistors

Resistors are used to limit current in circuits.

The three main considerations when choosing a resistor are:

- **Tolerance:** This gives a minimum and maximum value either side of the nominal value. A 50 ohm resistor could have a nominal value of 50 ohms with a tolerance of ±10%. Therefore it could have a value of 45 to 55 ohms.
- **Power rating:** This is the maximum power that can be developed in a resistor without causing damage to it.
- **Stability:** This is the ability of a resistor to maintain the same resistance regardless of temperature or age.

Resistors can be **fixed** or **variable** and there are various types. The main types of resistor are shown in the table below, with their properties.

	Carbon composition	Carbon film	Metal oxide	Wirewound
Maximum value	20 MΩ	10 MΩ	100 MΩ	270 MΩ
Tolerance	±10%	±5%	±2%	±5%
Power rating	0.125–1 W	0.25–2 W	0.5 W	2.5 W
Stability	Poor	Good	Very good	Very good
Use	General	General	Accurate work	Low values

The value of a resistor is coded in a special set of coloured bands printed on the resistor's surface. There are four bands on each resistor. Each coloured band represents a number.

The example below shows the bands on a resistor.

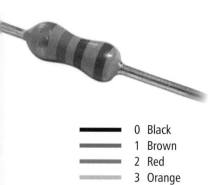

0	Black
1	Brown
2	Red
3	Orange
4	Yellow
5	Green
6	Blue
7	Violet
8	Grey
9	White

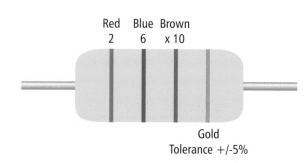

Red Blue Brown
2 6 x 10

Gold
Tolerance +/-5%

The first line is red (2) and the second line is blue (6). Total 26.

The third line is the multiplier. This gives the number of zeros after the number. Brown 1, Red 2, etc. Brown adds one zero to 26 which gives an answer of 260 ohms.

The fourth band gives the tolerance: gold ±5%, silver ±10% and no band means that there is no tolerance.

This resistor has a value of 260 ohms ±5%. So the top value is 273 ohms and the bottom value is 247 ohms.

What are the resistance values of the following resistors?

Think
IT THROUGH

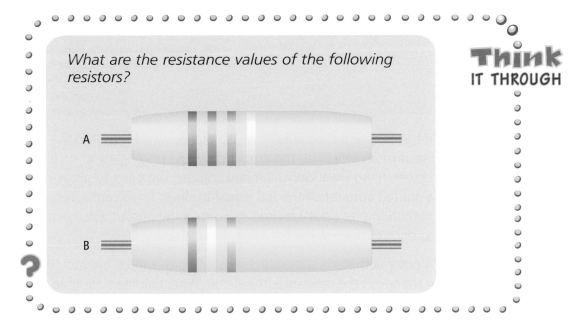

Solutions

Resistor A: 6500 ohms ±5%. Top value 6825 ohms, bottom value 6175 ohms.

Resistor B: 540,000,000 ohms with no tolerance.

Selecting thermistors

A **thermistor** is a resistor whose resistance varies with temperature. Thermistors are used in heaters, sensors and for current control. There are many different types of thermistors so choosing the correct one can be difficult.

Many suppliers produce easy-to-read **flow charts** which help customers find the right component for their application.

THE JARGON DRAGON

flow chart – a special diagram that can help direct the reader to a product, service or problem

Selecting potentiometers

Sometimes only a some of the **output voltage** from a signal source is needed.

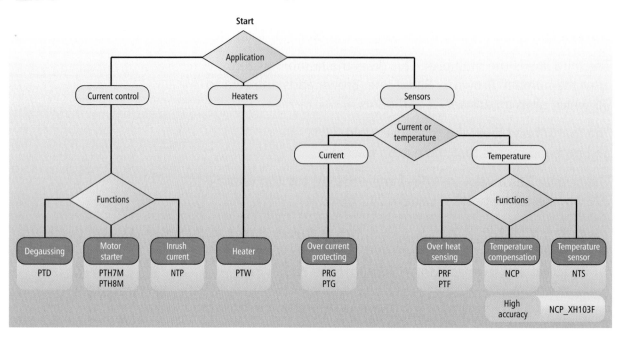

Example of a flow chart produced by a supplier to help customers find the right component

If the full output voltage from CD player was driven into the input of an amplifier, the amplifier would play nearly at its full power. If we want only a limited volume, we need to allow only a limited amount of the full signal through to the amplifier. A potentiometer ('pot') is used to control the output voltage and so control the volume.

A potentiometer is a modified resistor which can be used to allow a change in the resistance in a circuit. Potentiometers are used in volume controls that rotate, such as on a car radio or sound system.

The flow chart at the top of the page opposite is a supplier's guide to finding the appropriate potentiometer.

Capacitors

Capacitors are used in electric circuits. They store electric charge. The amount of charge they store is measured as capacitance (C). Capacitance is measured in farads, F.

If a capacitor stores one coulomb of charge when a potential difference (PD) of one volt is applied across the capacitor, then its capacitance is one farad.

There are two main factors that need to be considered when choosing capacitors:

- **working voltage:** the maximum working voltage that a capacitor can withstand
- **leakage current:** the loss of charge through the dielectric (the material between the plates of a capacitor).

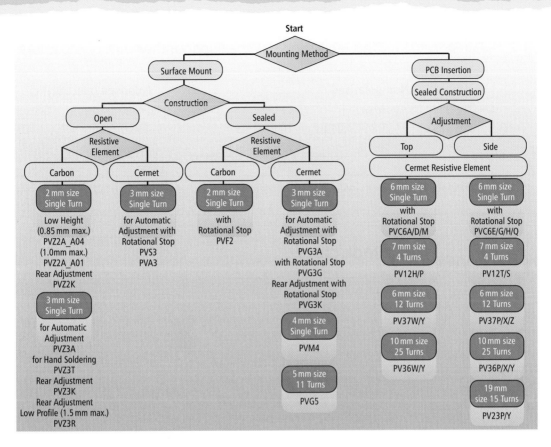

Capacitors are categorised as follows:

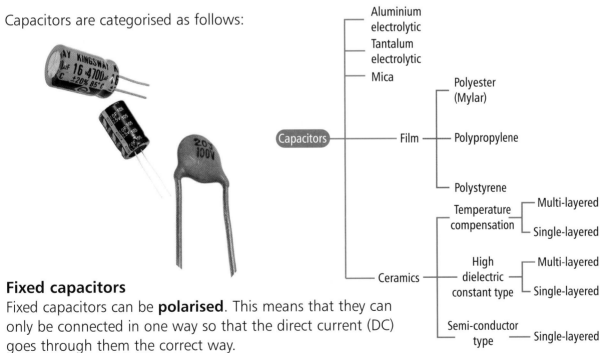

Fixed capacitors

Fixed capacitors can be **polarised**. This means that they can only be connected in one way so that the direct current (DC) goes through them the correct way.

Non-polarised capacitors can be connected either way.

Variable capacitors

These are generally used to tune radio receivers as their value can be varied.

This table shows all the important features of capacitors.

	Non-polarised			Polarised (electrolytic)	
	Polyester	Mica	Ceramic	Aluminium	Tantalum
Values	0.01–10 μF	1 pF– 0.01 μF	10 pF–1 μF	1–100,000 μF	0.1–100 μF
Tolerance	±20%	±1%	−25 to +50%	−10 to +50%	±20%
Leakage	Small	Small	Small	Large	Small
Use	General	High frequency	Decoupling	Low frequency	Low voltage

Note:

1 μF (1 microfarad) = one millionth of a farad (10^{-6} F)

1 pF (1 picofarad) = one millionth of a microfarad (10^{-12} F)

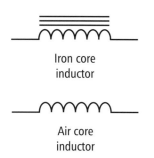

Iron core inductor

Air core inductor

THE JARGON DRAGON

core – the centre of an inductor. It is generally iron, or where there is nothing in the centre of the inductor, air

Inductors

An **inductor** can also be called a coil simply because it is a coil of wire.

There are several types of inductor and they are classified according to the material at the **core** of the inductor. The core is the centre of the inductor. An inductor is made by forming a coil of wire around the core.

There are two main types of inductor core: iron and air.

In an iron core inductor the wire is wrapped around the iron core. In the symbol the core is shown as three bars. In an air core inductor there is no material at the core but there is usually a thin tube of material, which is a non-magnetic, that the wire is wrapped around.

The main factors that affect inductance are:

- the diameter of the coil
- the length of the coil
- the core material
- the number of windings in the coil
- the number of turns in the coil.

Diodes

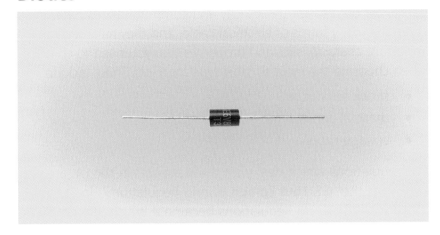

A **diode** is a semi-conducting device, which allows electricity to flow in one direction but not the other.

There are several different types of diode. It is important to consider what the diode will be used for in order to choose the correct type. Diodes convert alternating current (AC) to direct current (DC). There are special terms used for each side of the diode. The positive side of the diode is called the anode which is made from p-type material The negative side is called the cathode and is made from n-type material. Current always flows from the anode to the cathode. This type of diode is known as a junction diode.

Heavy and light currents use different types of diode, so this factor needs to be considered when deciding what type of diode to use. The speed of the signal also needs to be taken into account. Another important consideration is the application of the diode. For example, when a number of diodes are connected in a particular way, alternating current can be converted to direct current.

LEDs (light-emitting diodes)
There are many types of LED. Some can give off light, and some can detect light. LEDs are used as indicators, usually small red or green lights, on many electrical products. They should never be used in circuits where more than 40 mA (milliamperes) or 2.2 V (volts) is reached.

Zener diodes
These diodes are silicon junction diodes which maintain a steady output voltage from a power supply.

Cables

Cables are used to carry electricity and communications data. There are many forms of electrical cable, and they have several important characteristics and properties that you need to consider when choosing the most suitable type for a job. These include:

- shock resistance
- thermal constraints
- voltage drop.

The regulations which deal with these aspects in the UK are known as the 'IEE 16th Edition Wiring Regulations'. In October 1992 the **IEE** Wiring Regulations became a British Standard, BS7671, giving them international recognition.

Choosing cable

Many cable suppliers show their products' specifications in tables like the one below. These tables help customers to find the best cable to use for a particular job.

CSA (mm^2)	Construction	Current rating	Radial thickness (mm) Insulation	Sheath	Nominal diameter	Weight kg / km
0.5	16 / 0.2 mm	3 amps	0.6	0.8	6.3	40
0.75	24 / 0.2 mm	6 amps	0.6	0.8	6.8	63
1.0	32 / 0.2 mm	10 amps	0.6	0.8	7.2	73
1.25	40 / 0.2 mm	13 amps	0.7	0.8	7.8	84
1.5	30 / 0.25 mm	15 amps	0.7	0.8	8.2	95
2.5	50 / 0.25 mm	20 amps	0.8	1.0	10.0	154

Example of a table showing electrical cable properties

The CSA (cross-sectional area) gives the thickness of the cable.

An excellent range of cables and their properties can be found at `http://www.cse-distributors.co.uk/cable/cableind.htm`

Cable manufacturers may give a written description of each cable to help customers understand their use.

Flexible data transmission cable

'These cables are employed for pulse-operated or AC circuits in control, measuring and signalling technology as well as in data processing and office technology. Twisting in pairs provides optimum crosstalk attainability'.

Specification:

- Fine strands of bare copper conductor
- Stranding acc. to VDE 0295, class 5
- PVC core insulation
- Core marking acc. to DIN 47100
- 2 cores twisted in pairs
- Pairs twisted in layers
- Film wrapping
- Overall screen of tinned copper wires
- PVC outer sheath grey, RAL 7001 or 7032
- Temperature range flexing: −5°C to +70°C; static: −30°C to +70°C

Highly flexible polyurethane cable

'This cable is for use in control circuits, measuring and also in power circuits, in the automation industry, in assembly lines and production lines and is suitable in drag chains where the cable is exposed to fast and abrupt movements.'

Specification:

- Super fine strands of bare copper wire
- Stranding acc. to VDE 0295, class 6
- PVC core insulation black with continuous white figure imprint
- Cores twisted in layers
- Earth conductor green/yellow in outer layer
- Tex tail wrapped
- Screen made of tinned copper wire
- PUR outer sheath – grey, RAL 7001
- Microbe- and hydrolysis-resistant, adhesion-free, flame retardant
- Bending radius 7.5 × Δ
- Temperature range flexing: −10°C to +70°C

Batteries

Batteries are used in many of the products that we use, for example radios, calculators, remote controls, cameras, compact disk players, camcorders, mobile phones, etc.

A **primary cell** or battery is a battery that cannot easily be recharged after one use, and it is discarded following discharge. Most primary cells use **electrolytes** that are contained within absorbent material or a **separator**. This means that there is no free or liquid electrolyte. They are therefore termed **dry cells**.

A **secondary cell** or battery is a battery that can be electrically recharged after use to its original pre-discharge condition. This is done by passing current through the circuit in the opposite direction to the current during discharge.

There are many types of batteries available. They are classified according to the chemicals that they are made from. Examples include:

- carbon zinc
- alkaline manganese dioxide zinc (cylindrical)
- alkaline manganese dioxide zinc (miniature)
- lithium manganese dioxide (Li/MnO$_2$)
- silver oxide zinc
- sealed lead acid.

Manufacturers produce tables of their products so customers can quickly compare the various types of battery. Battery sizes are in the format of AA, AAA, C, D, F and N batteries.

Here are the main characteristics of a range of batteries.

Name	Size	Capacity (mAh)	Voltage (nominal)	Weight (g)	Diameter (max mm)	Height (max mm)
3-312	AAA	1250	1.5	11.5	10.29	42.82
3-312I	AAA	1250	1.5	11.5	10.50	43.27
3-315	AA	2850	1.5	22.4	13.99	47.96
3-315I	AA	2850	1.5	22.4	14.3	48.41

Motors

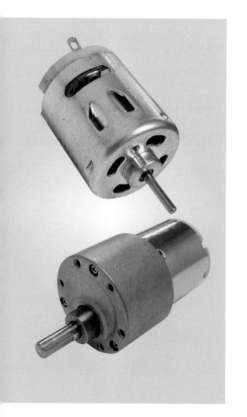

AC motors

Information can be found about any motor by checking its nameplate.The most commonly used motor is the three-phase AC induction motor. Its advantages are:

- simple design
- low cost
- reliable operation
- replacements are easily found
- variety of mounting styles
- many different environmental enclosures.

The AC motor has a simple design. It consists of a series of three windings in the exterior or **stator** section with a simple rotating section, or **rotor**. The interaction of the magnetic fields between the stator and rotor results in mechanical rotation.

The speed of an AC motor depends on only three variables:

- the fixed number of winding sets, called **poles**, built into the motor, which determine the motor's speed
- the **frequency** of the AC line voltage. Variable speed drives change this frequency to change the speed of the motor
- the amount of **torque** loading on the motor, which causes **slip**.

DC motors

The brushed DC motor is one of the earliest motor designs. This motor is usually chosen where variable speeds are required. Its advantages are:

- the design is easy to understand
- speed is easily controlled
- torque is easily controlled
- it is cheap and simple to design.

DC motor design is quite simple. A permanent magnetic field is created in the stator by either permanent magnets or electro-magnetic windings.

Audible alarms (can be called AAs or buzzers)

Audible alarms are designed to give an output that can be heard. There are many types available. Some commonly used types of audible alarm are detailed below. There are many excellent websites giving product ranges – try this site:

`http://www.thomasregister.com/olc/55751051/`
`acoustic.htm`

Magnetic audible alarms use a coil of wire. When a switch is pressed an electric current flows through the coil. This causes the coil to create a magnetic field. The magnetic field pulls a piece of metal towards it, the metal strikes another piece of metal and this causes a sound to be made.

A magnetic buzzer

A **piezo buzzer** uses a transistor circuit to supply a varying voltage to a piezo-electric disk. The disk flexes and vibrates when the varying voltage is applied to it, which generates a sound.

A piezo buzzer

Mechanical audible alarms

Manufacturers' guides can help when choosing alarms. Audible alarms are classified by sound output in decibels (dBA/cm). The higher the number the louder the sound.

Think
IT THROUGH

Think of as many products as you can that use audible alarms.

Make a list and put them in order of how loud you think they are.

The chart below shows a supplier's range of mechanical buzzers and their sound output.

Model	Min. sound output (dBA/cm)	Model	Min. sound output (dBA/cm)
20S1015	70/20	66S1120	75/20
63S1015	70/20	68S1120	75/20
66S1015	70/20	20S1240	75/20
68S1015	70/20	63S1240	75/20
20S1030	70/20	66S1240	75/20
63S1030	70/20	68S1240	75/20
66S1030	70/20	34S1060	95/10

Lamps

Lamps give out light when an electric current passes through them.

Illumination lamp

They have a thin wire **filament** which becomes very hot when a current passes through it. This causes the lamp to glow. The filament material needs to have a high melting point; materials such as tungsten are used for filaments.

Indicator lamp

Filament lamps need to be replaced often as they eventually burn out. Care must be taken when handling them as the filament can easily be broken.

Some lamps are used to provide continuous illumination and others need only be bright enough to be used as an indicator, such as on machine controls. There are symbols for each of these types.

Selecting the most appropriate lamp

The three most important points to consider are:

- **voltage rating:** the supply voltage for normal use
- **power or current rating:** smaller lamps are usually rated by current.
- **lamp type:** lamps can be different shapes and have different ways of fastening into a lamp holder.

The voltage and power (or current) ratings are usually printed or embossed on the body of the lamp. Larger lamps have the ratings printed on the glass.

Miniature Edison screw (MES)

This is the standard small lamp. The bulb diameter is usually about 10 mm.

Miniature centre contact (MCC)

This type of lamp fits into the lamp holder in the same way as most household lamps. This is known as a bayonet-style fitting. MCCs are similar in size to MESs.

Pre-focus

This type of lamp is generally used in torches. The lamp is held in place by the components of the torch in which it is fitted. There are not many lamp holders suitable for this type of lamp so they are generally not used in circuits.

Small bayonet cap (SBC)

This lamp type has a bayonet-style fitting. It also has a high power rating and the bulb is large.

Reviewing existing products and their materials

Looking at existing products and the materials they are made from will give you a good idea of the suitability of different materials for different jobs.

Metal	Typical components
Grey iron	Engine blocks, brakes, gears, machine bases
Zinc	Door handles, carburettors, small technical parts such as gears
Brass	Ornaments, pipe fittings, valves
Aluminium	Pistons ,manifolds, aircraft components
Alloy steel	Gas turbine housings, car wheels
Mild steel	Car bodies, tools, fasteners
Copper	Pumps, valves, marine components

Form of supply

Materials can be supplied in many forms.

Materials that can be extruded or rolled can be supplied in long bars. Typical sections are shown on the following page. As you can see, materials can come in a wide range of shapes.

Not all suppliers supply all materials. It is important to look at suppliers' catalogues and websites for information to help you identify the best material to use for the products you are going to make.

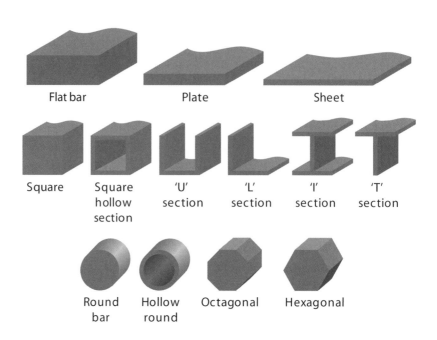

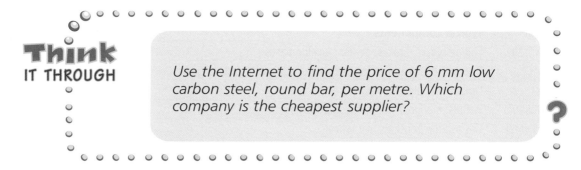

Forms of supply of materials

Size of material

Bars and sections are manufactured in different sizes. It is important to consider the size of the bar, as you may need to remove material from it. You will want the bar that is nearest to the size you require.

Think IT THROUGH

Use the Internet to find the price of 6 mm low carbon steel, round bar, per metre. Which company is the cheapest supplier?

If you require material of an unusual size, it is likely that it will not be available, or it will have to be manufactured specially for you. This will be very expensive and is only done in special cases.

Polymers are supplied in grains which are fed into various moulding machines, or as sheet which is formed into products using vacuum-forming machines.

Polymers are coloured, and suppliers are able to supply polymers in an almost endless range of colours. There is also a range of different colouring effects available, such as marbling or speckling. Designers of products need to work closely with suppliers to make sure they have the best choice of materials available.

Handling materials

Materials can be difficult to handle. Machines are often used to lift and move material. You should consider the following:

- How big is the material to be used?
- What is the density or weight of the material?
- How sharp are the edges?

Bars, sheets and sections of material can be extremely heavy and awkward to carry. Consideration must be given to what will be a suitable size for the material to be supplied in.

- Does the company have lifting equipment that is able to handle the material?
- Does the company have the facility for cutting or sawing the material?

The weight and density of material

Relative density gives a guide to a material's weight. Water has a relative density of 1. Relative density is how many more times heavier a material is than water. Aluminium has a relative density of 2.7 and so is 2.7 times heavier than water.

The table and chart shows a range of materials and their relative densities.

Metal	Relative density
Aluminium	2.7
Zinc	6.8
Grey iron	7.4
Alloy steel	7.8
Mild steel	7.8
Brass	8.5
Copper	8.9

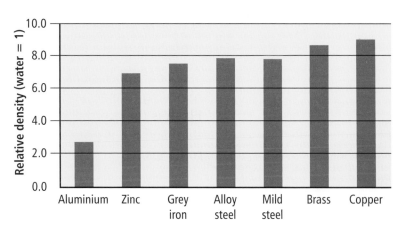

Relative density of engineering materials

It is difficult to say that one piece of material is heavier than another because one piece may be bigger than the other. Density is used to give a measure of how heavy a material would be if it were a given size. It is based upon how much the material would weigh if it were a cubic metre in size. That is equivalent to a box, one metre long, one metre wide and one metre tall.

Water has a weight of 1000 kg per cubic metre.

Mild steel has a relative density of 7.8. This means that it is 7.8 times heavier than water.

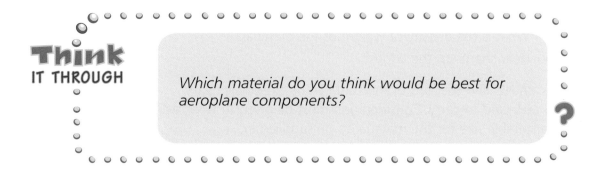

Think
IT THROUGH

Which material do you think would be best for aeroplane components?

Safety in handling materials

Sawn materials
Sawn materials may have hidden dangers due to the way that they have been manufactured. Bars are cut to length with powerful industrial saws. This leaves sharp edges. Gloves should be worn when handling large pieces of sawn metals.

Castings
Castings can be hot for a long time, especially if they are large components. In foundries, castings should always be treated with care as heat cannot be seen. Always use leather heat-protection gloves in this type of environment.

Secondary operations are carried out on castings such as removing runners and risers. This can leave very sharp edges. **Flash** is the overspill of material in the mould. This can be very thin, and if it is not removed from the mould it can cause cuts to hands.

Flame-cut material
Plate is often cut using a flame cutter. This can leave very sharp edges. Plate is generally heavy and the combination of sharp edges and weight can make it dangerous to handle. Always wear gloves when handling flame-cut plate.

Cost of materials

Different materials cost different amounts. Suppliers who compete with each other may offer the same material at different prices, so there is no definite answer as to how much materials cost.

Engineers look at different suppliers in the same way that you look in different shops to buy a particular product. Cost can be generalised to give an idea of the relative cost of materials. Mild steel is used so much that it can be a useful gauge for comparing the costs of other materials. If we say that mild steel has a cost value of 1, then a material with a value of 2 costs twice as much.

Metal	Bar form	Sheet form
Mild steel	1.00	1.00
Brass	6.50	4.00
Aluminium	8.80	6.70
Stainless steel	9.70	6.30
Copper	12.00	7.00
Titanium	24.00	20.00

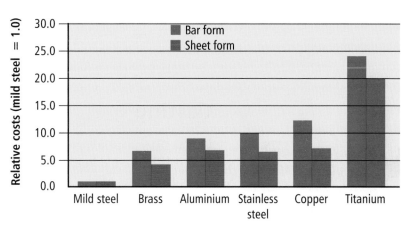

Relative cost of engineering materials

The form of the material can affect the cost. Sheet form is often cheaper than bar form.

We can use this chart to help us to work out the costs of materials for a component.

Choosing materials

When engineers choose the material for their products, they have to take into account all the various points such as cost, process capability, weight, form of supply and so on. There are often conflict between designers, managers and engineers because designers want the product to look the best it can, managers would like to make the product in the most cost-effective way and engineers want the products to function as best it can. When deciding on a material for a product, all these considerations must be taken into account.

Processes

There are many ways in which products can be made. The materials that are used can be affected by the manufacturing process, and the manufacturing process can be affected by the materials used to make the product.

It is therefore important to understand the different manufacturing processes that are available.

Manufacturing processes fall into four main categories:

- casting
- cutting
- forming
- joining.

Casting

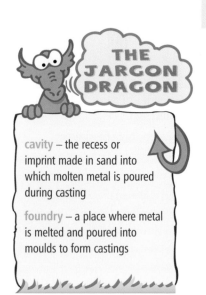

cavity – the recess or imprint made in sand into which molten metal is poured during casting

foundry – a place where metal is melted and poured into moulds to form castings

In this process material is heated to above its melting point so that it becomes a liquid. The liquid is poured into a mould and takes the form of the hollow area known as the **cavity**. When the material cools it is removed from the cavity and is the same shape as the cavity.

Sand casting

Sand casting is a process of moulding components from molten metal. This takes place in a **foundry**.

Sand casting uses what is known as a single-use mould. First a wooden version of the product is made. This is known as a **pattern**. The pattern is placed in a box known as a **drag**. A special sand is poured into the drag until it is full and the pattern is covered. The pattern is removed and a box known as a **cope** is placed on top of the drag. The cope and the drag are then assembled together.

Molten material is poured into the cavity through a special hole called a **sprue**. When the cavity is filled the excess material flows up an escape hole known as a **riser**. The material is left to cool and the sand is broken away. Each time a new product is to be moulded a new cavity has to be produced.

This process is best for making products in small batches or single components as it saves the cost of making a steel mould.

In a foundry, most materials can be used as the equipment is able to heat metal to a very high temperature. Steel, for example, can be melted in a foundry. Sand casting can be used in smaller workshops but there is usually a limit to the temperatures that can be reached and therefore the range of metals that can be used.

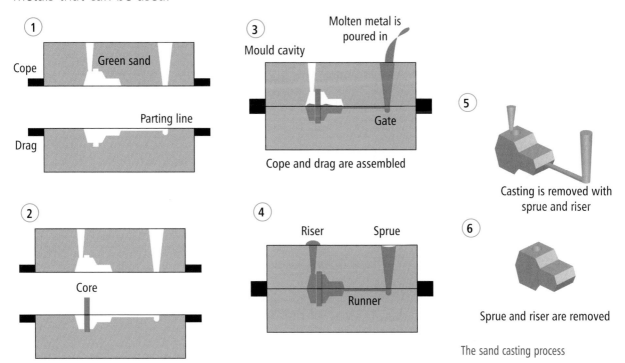

The sand casting process

Die casting

Gravity die casting

If a large number of products are to be moulded, then it is usual to produce a **die** in which to pour the molten metal.

A die is produced in two halves which can brought together so that the molten metal can be poured into the cavity. When the cavity is full the component is left to set. The two halves of the die are then separated and the component removed.

Gravity die casting uses the force of gravity to force the molten material into the cavity.

Pressure die casting

In this process, pressure is used to force the molten material into the cavity. It generally gives more accurate and more detailed castings than gravity die casting.

Metal	Castability
Mild steel	Fair
Alloy steel	Fair
Copper	Fair / Good
Aluminium	Good / Excellent
Brass	Excellent
Zinc	Excellent
Grey iron	Excellent

Castability of engineering materials

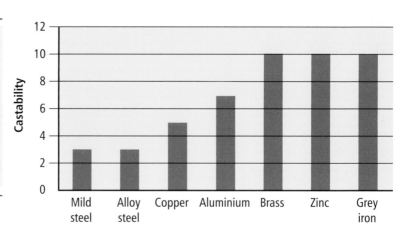

The table and graph above show how well materials can be cast. This property is known as **castability**.

Cutting

A cutting process starts with a piece of material known as a **billet**. Material is removed from the billet to form a product. This can be done by hand, such as sawing and filing, or using machines. Machines can use single-point tools, such as a turning tool on a centre lathe, or multiple-point tools such as a milling cutter.

Different cutting tools are used to create different shapes in the material. Some of these are shown below.

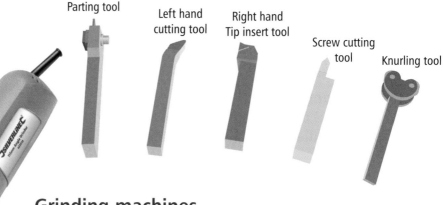

Parting tool
Left hand cutting tool
Right hand Tip insert tool
Screw cutting tool
Knurling tool

Grinding machines
Sometimes referred to as '**grinders**', grinding machines have several designs. They use grinding wheels made from extremely hard ceramic materials that are bonded together.

Bench grinders
When a bench grinder is being used, the work piece is usually held by hand. Material is removed from the work piece by moving it across the rotating grinding wheel.

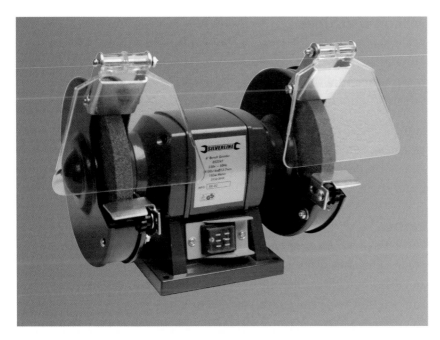

A grinding wheel

Bench grinder

Grinders are best at grinding hard materials, as soft material can block up the grit that forms the grinding wheel.

This process is used to remove hardened steel from work pieces such as chisel points, scribers and centre punches. It is also used to sharpen lathe cutting tools. The cutting tool can be rested upon adjustably angled plates that give the correct form to the turning tool.

Grinding machines such as **surface grinders** and **cylindrical grinders** are extremely accurate but can remove only small amounts of material. Usually large amounts of material are removed first using processes such as milling and turning before using these grinders.

Surface grinders
Surface grinders are generally used to form very flat surfaces on materials. However, the grinding wheel can be shaped using an industrial diamond to produce a profile along the work piece. They are precision machines which can be accurate to 0.001 mm.

Cylindrical grinders
A cylindrical grinder is used to produce components that are cylindrical in shape. Cylindrical grinders remove material very slowly, so it is usual to produce the component on a centre lathe first and to then use a cylindrical grinder to finish the diameter to the correct size.

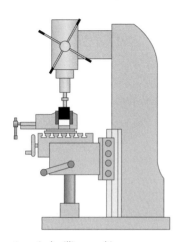

A vertical milling machine

Milling machines

Milling machines use cutting tools with more than one point known as multi-point cutting tools. They can remove material more efficiently than lathes as there are more 'teeth' with which to cut the work piece.

Light and soft alloys can be cut using milling machines, as well as carbon steels. As steels become harder they can approach the same hardness as the cutting tool. This will make the cutting tool blunt.

Lathes

Lathes use single-point tools to remove metal from the work piece. They are able to cut light alloys and heavy alloys as well as carbon steels easily. Harder steel can be cut, but for this a tool made of special materials is required.

Lathes can be accurate to 0.01 mm, but there are limits on the size of the material that they can cut.

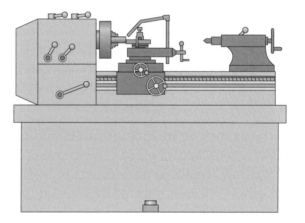

A lathe

Machinability gives an indication of how easy a material is to cut. The table shows a range of materials with their machinability on a scale of 0 to 10.

Metal	Machinability
Alloy steel	Fair
Mild steel	Fair / Good
Grey iron	Good
Aluminium	Good / Excellent
Copper	Good / Excellent
Brass	Excellent
Zinc	Excellent

Machinability of engineering materials

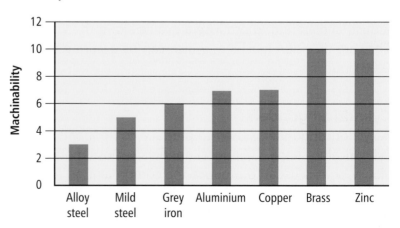

Forming

Sheet forming

Sheet forming involves the deformation of thin sheets of material. The sheets can be up to 3 mm thick, but the thicker the material the more power is required and there is a greater chance that the material will fracture or break. The material should therefore be **ductile** and not brittle as it will be stretched and compressed. Car bodies, filing cabinets and domestic appliance covers are all manufactured in this way.

A **tensile test** shows how a material reacts when it is stretched. The stretching is called **elongation**. Materials with good elongation can be formed easily and so have good **formability**.

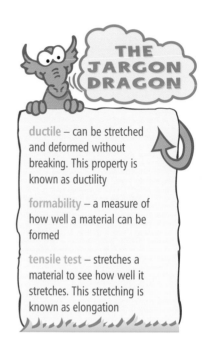

THE JARGON DRAGON

ductile – can be stretched and deformed without breaking. This property is known as ductility

formability – a measure of how well a material can be formed

tensile test – stretches a material to see how well it stretches. This stretching is known as elongation

Test specimen is stretched

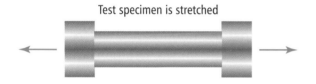

The test specimen stretches and breaks

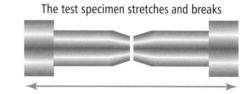

Metal	Elongation %
Grey iron	5
Zinc	15
Brass	20
Aluminium	30
Alloy steel	30
Mild steel	40
Copper	55

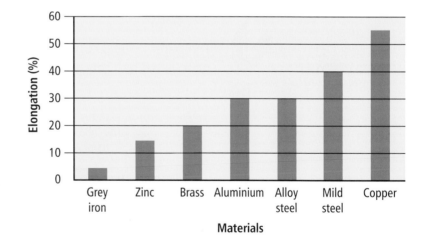

Elongation of engineering materials

Bulk forming
Forging

During this process, large blocks of material are compressed into a cavity that is the shape of the product. The material used needs to be malleable so that it can deform into its new shape. Steel is often used, but it needs to be heated to red hot temperatures before it becomes malleable. Lead can also be used.

Spanners, G-clamps and coins are pressed using this process.

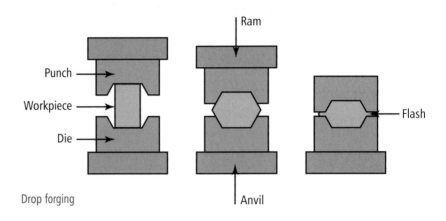

Drop forging

Joining

Non-fusion welding

This is a type of welding in which the **parent metal** does not melt. A **filler material** is heated and melts, and this bonds to the surface of each of the two pieces of metal to be joined together.

There are several processes that use non-fusion welding. The principle is the same for each of them. The two parts of material to be joined are brought together. Heat is applied to the parent metal but does not melt it, and heat is applied to melt filler material which then flows into the gap between the two parts.

The parent metal surfaces must be clean. **Flux** is used to remove oxides. This process is known as 'wetting the surface'.

THE JARGON DRAGON

filler material – the material that is melted to fill the gap between two pieces of metal that are to be joined together during non-fusion welding

parent metal – the main pieces of metal that are to be bonded together during welding

weld bead – the joint created that holds two materials together during fusion welding

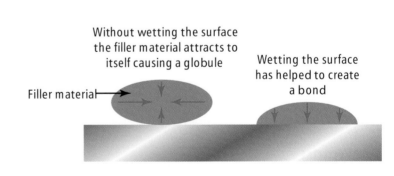

'Wetting the surface' during non-fusion welding

Non-fusion welding processes include the following.

Soldering

This process is used where the temperature of the filler rod falls below 450°C. It is used to join electrical components and thin sheet.

Brazing

Brazing is used for material that requires a filler rod that melts above 450°C. It is used to join dissimilar metals.

Surfaces are cleaned and wetted

Manual metal arc welding

This is the most widely used method of welding. In this method the filler rod is also used to carry the **arc** (electricity) to the joint. The rod is covered in a hard flux. As the flux burns it gives off a protective shield of gases around the weld.

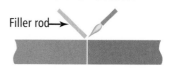

Heat is applied and filler material is introduced

Filler rod

Plasma arc welding

This is a recent development. A gas such as argon, air or CO_2 is passed through a constricted electrical arc. This causes the gas to **ionise** and create a mixture of electrons and positively charged particles known as **plasma**. The high temperature created in this method can also be used for cutting, but the pressure needs to be much greater.

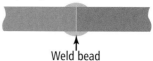

Filler material runs into the gap by capillary action to form a bond

Weld bead

The process of non-fusion welding

Oxy-acetylene (gas) welding

The flame created by the two gases oxygen and acetylene is the hottest known flame.

The development of gas shielding in arc welding has meant that oxy-acetylene welding is used less than it used to be, though it is still used for general maintenance work and small workshop jobs.

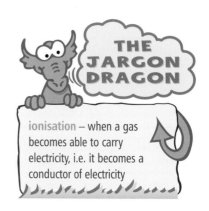

THE JARGON DRAGON

ionisation – when a gas becomes able to carry electricity, i.e. it becomes a conductor of electricity

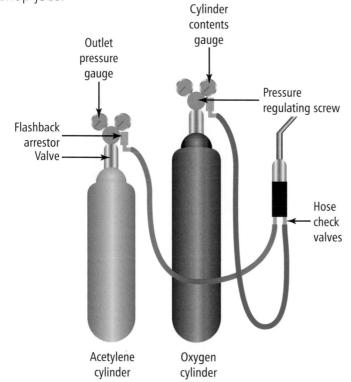

Cylinder contents gauge

Outlet pressure gauge

Pressure regulating screw

Flashback arrestor Valve

Hose check valves

Acetylene cylinder

Oxygen cylinder

homogeneous – a material that is made up of only one material, not a mixture

weldability – the ability of materials to be fusion welded

Fusion welding

In fusion welding the parent metal is melted and forms the filler. This means that the weld bead is the same material as the parent metal. The two pieces of metal fuse together and become **homogeneous**.

Parent metal is chamfered to assist welding process

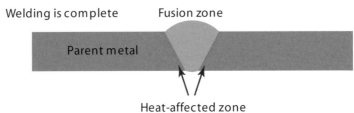

Fusion welding

MIG welding (metal inert gas welding)

MIG welding uses a filler material that is wound around a drum known as a spool. The filler is fed automatically into the gap by a hand-operated trigger.

MIG welding is seen as a clean process because of the lack of flux. It is used in a wide range of applications such as ship building, car manufacturing and the aircraft industry.

The table and chart shows the **weldability** of a variety of engineering materials.

Metal	Weldability
Grey iron	Specialist welding
Zinc	Specialist welding
Aluminium	Fair
Copper	Fair
Brass	Good
Alloy steel	Excellent
Mild steel	Excellent

Weldability of engineering materials

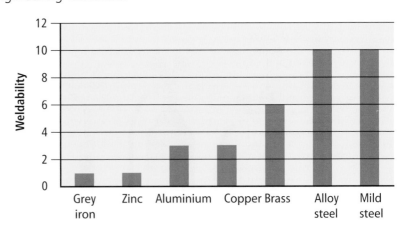

Adhesives

Adhesives are chemical liquids that bond surfaces of materials together. They are commonly referred to as glue. Heat may be used in some cases to aid the setting of the adhesive. Most adhesives are designed to bond particular materials together, so it is important to consider which type of adhesives can be used on the materials you are working with.

Adhesives perform better the larger the surface areas that are in contact. If components are joined end to end with adhesives, they may break due to the bending stresses at the joint.

Some of the most powerful adhesives are supplied as two materials which are mixed together just prior to use. A chemical reaction between them creates the adhesive. The chemical used to trigger the adhesive effect is known as a **catalyst**.

Types of adhesive
Rubber-based cement
This is used to join light plastics. The adhesive is applied to both parts to be joined and left to dry.

Superglue
Superglue is a very strong multipurpose glue which will join many materials. It comes in liquid form in a tube and dries very quickly. Users must be careful to avoid contact with the skin and eyes.

Epoxy resin glue
This is supplied in two parts which are mixed together just before being used. It can join a wide variety of materials. The bond is very strong and dries in about five minutes.

Examples of adhesives

Using processes

Joining and assembly

There are three main ways in which materials can be joined permanently using heat. These are **soldering**, **brazing** and **welding**.

Soldering
This process can be used to join two pieces of metal together.

It involves heating a material, but not to melting point, and heating a **filler material** (solder) until it is at melting point. The two pieces of metal are held in place, and as the **solder** melts it flows over the two pieces of metal. When the solder solidifies, a single component is produced. The original material does not melt, but the freezing solder bonds to the surface of the **parent material** and joins the two pieces of metal together.

Soldering does not produce a strong joint, so it is generally used only for joining parts together end on. This is known as a **butt joint**. It is used for joining copper wires only where no force is applied trying to pull the wires apart.

All soldered joints must first be prepared.

Types of solder

Solder is a mixture of tin (chemical symbol Sn) and lead (chemical symbol Pb), but they are not always mixed in the same ratio. Different types of solder have different ratios of tin to lead.

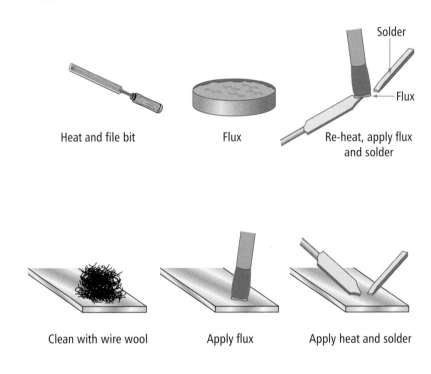

Heat and file bit   Flux   Re-heat, apply flux and solder

Clean with wire wool   Apply flux   Apply heat and solder

It is important that, when you are soldering a joint, the most appropriate solder is used. The British Standard specification EN29453 provides details of a number of different solders, but there are many more types of solder than are listed here.

Remember that the melting point of the solder should be lower than that of the parent material, otherwise the parent material could melt or be damaged before the solder melts.

BS specification EN29453

Tin (Sn) %	Lead (Pb) %	Other	Freezing range (°C)	Uses
63	37		183	'Tinmans solder' – used for joining, in printed circuits
50	50		183–215	General sheet metal work
3	96	copper 1%	230–240	Joining copper pipes

Joint preparation

Preparation of the joint is extremely important in soldering. The two pieces of metal must first be formed into the shape of the joint, such as a partition joint, corner joint or self-secured folded joint.

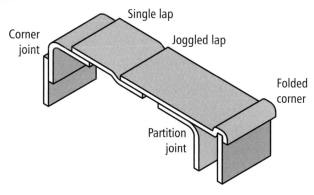

Types of soldered joint

The surfaces that are to be soldered need to be extremely clean. Cleaning is usually done using a special chemical known as a **flux**, which even removes the oxides on the surface of the parent material.

Now the parent material is ready to be heated with the soldering iron or flame. The heat is spread around the joint. When solder is applied to the copper bit of the soldering iron, it melts and runs onto the parent material. The flowing solder runs into the small gap between the two surfaces to be joined together by capillary action. When the solder freezes the joint is secure. The joint must be fully soldered so that the joint is waterproof.

Flux

Joints should be cleaned with an abrasive material such as wire wool, but a flux must always be applied too. When metals are left in the atmosphere they react with oxygen in the air, which creates a coating known as an oxide on the metal. Fluxes are needed to remove this oxide.

Fluxes fall into two categories: **corrosive** and **non-corrosive**. An example of a corrosive flux is hydrochloric acid, which is used for the solder of zinc and zinc-covered steel (galvanised steel). An example of a non-corrosive flux is natural resin dissolved in methylated spirit. This is used for electrical joints and containers.

Brazing

Brazing is basically the same process as soldering, but the fillers used have melting temperatures higher than 450°C. The brazing process can be used on most metals, including special engineering metals.

Brazing gives a much stronger and tougher joint than soldering. When deciding whether to braze or solder, brazing will give the stronger joint, but as the melting point of the fillers used is 450°C, the melting point of the parent material needs to be even higher. This is an important consideration. It is also important to realise that it costs more, and it takes more time to heat up the filler material.

Brazing can be done at a wide range of temperatures, and this means that a different flux needs to be used with each filler material. Borax-type fluxes are used with ferrous materials, but silver soldering uses lower temperatures so a fluoride flux can be used.

Brazing alloys

BS specification 1845

Copper (Cu), %	Zinc (Zn), %	Others	Freezing range (°C)	Uses
60	39.7	silicon – 0.3%	875–895	Copper, steels, nickel alloys
28	–	Ag – 72	780	
15	16	Ag – 50 Cd – 19	620–640	Copper, copper alloys, silver solder

Heat treatment of steels

Steel is an alloy of iron with a small amount of carbon. Steel has certain properties, including hardness, toughness, strength and ductility. These properties can be changed by heating the steel to a given temperature and then cooling it either quickly or slowly. By using different temperatures and cooling methods, engineers can transform the original steel they buy from the supplier into steel with very different properties.

When heat treating steel, there are three important points to consider:

- the carbon content of the steel
- the heating temperature
- the cooling method.

heat treatment – the process of changing the properties of metals

Carbon content

The amount of carbon in the steel will affect the changes that take place during heat treatment. If there is no carbon at all, or even if there is a small amount, such as less than 0.3%, then the properties of the steel will not be changed. As the amount of carbon increases then so does the effect of heat treatment.

Heating temperature

There is no set temperature to heat steel up to in order to harden it. This temperature depends on the carbon content.

The graph shows how the upper critical point varies with carbon content. In order to harden the steel, the steel must be heated to between 30°C and 50°C above this upper critical point.

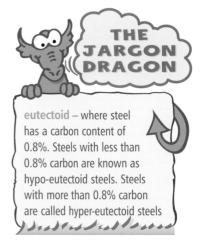

eutectoid – where steel has a carbon content of 0.8%. Steels with less than 0.8% carbon are known as hypo-eutectoid steels. Steels with more than 0.8% carbon are called hyper-eutectoid steels

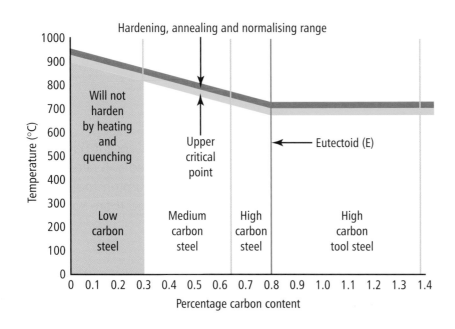

Iron–carbon diagram (Adapted from Higgins, RA 'Materials for the Engineering Technician', 3rd Edition. Butterworth Heinemann)

It can be seen that if we know the carbon content, for example 0.5% carbon, then we can estimate from the graph the upper critical point, in this case 780°C. To harden the steel, the steel must be heated to 30–50°C above this, so that would be 810°C–830°C.

To what temperature would steel of 0.4% carbon need to be heated, in order to harden the steel?

Cooling the steel

After the steel has been heated, it must be cooled. This can be done slowly by letting the steel cool naturally in air, or quickly by quenching it in water. Sometimes the steel can be cooled in oil, which cools it quickly, but not as quickly as water. This has the advantage of not cracking or distorting the steel, which can sometimes occur during water cooling. In some cases even cooling in air is too fast, so the steel is cooled in a furnace which is gradually turned down.

There are several types of heat treatment processes:

- hardening
- tempering
- annealing
- normalising.

Hardening

This process makes the steel harder so that it will scratch or dent other steels that have not been heat-treated. It can be used to harden scribers, centre punches and vice jaws. The steel cannot be hardened too much as it may become too brittle for some applications and crack or break.

Tempering

This process reduces the hardness of the steel once it has been hardened. Some tools that have been hardened will need to be tempered only a small amount, for example turning tools. Other products, such as springs, need to be tempered more so they are still hard but have flexibility.

Annealing

When steels and other metals are worked upon in the workshop, they can be bent, hammered, heated or welded.

Temperature °C	Colour Chart	Colour Description	Tools
300		Blue	Springs
290		Deep blue	Screwdrivers
280		Dark purple	Wood chisels
270		Purple	Axes
260		Purple brown	Knife blades
250		Brown	Punches, dies, woodworking tools
240		Dark straw	Drills, end mills, slot drills
230		Straw	Hammers, slotting tools
220		Pale straw	Turning tools

Examples of tools and their tempering temperature

These processes cause the metals to become hard. This is known as work hardening. In order to take the metal back to its original properties, it is heated and left to cool very slowly.

Normalising

Some processes, such as forging, heat the metal, and this changes the properties of the metal. As the temperature increases so does the size of the **grains** within the metal. In order to make the grains the original size, and take the metal back to its original properties, the metal is normalised by heating it and leaving it to cool in air.

THE JARGON DRAGON

grains – the name given to the particles that steel is made of

A summary of heat treatment processes

Process	Heat treatment	Cooling method
Hardening Makes the steel hard	30°C–50°C above upper critical point	Quench in water or oil
Tempering Lowers the hardness of the steel after hardening	This depends on products to be treated (use the tempering chart)	Cool in water or air
Annealing Brings work-hardened metal back to original properties	30°C–50°C above upper critical point	Leave to cool in sand
Normalising Gives heated metals original grain size and properties	30°C–50°C above upper critical point	Cool in air

Measuring temperature of heated steel

When steel is being heated, the colour of the metal changes as the temperature changes. Between 0°C and 450°C an oxide on the surface of the steel changes colour. After this the steel becomes so hot that it starts to glow in different shades of red and white. We can use these different colours as a guide to how hot the steel has become.

The temperatures required for hardening, annealing and normalising are above 723°C. The steel becomes red hot. The temperature can be estimated by the shade of the red.

The following chart shows how the redness relates to the temperature of the steel.

Temperature °C	Colour chart	Colour description
1200		White
1100		Light yellow
1050		Yellow
980		Light orange
930		Orange
870		Light red
810		Light cherry
760		Cherry
700		Dark cherry
650		Blood red
600		Brown red

How hot do you think these pieces of steel are, using the chart above?

Etching

Etching is a chemical process that removes material through the reaction of a chemical with the material being etched.

A mask is placed over the material to be processed, which protects parts of the material and leaves other parts exposed. The chemical is then applied. When the mask is removed the material under the mask will not have been eroded and will be left upstanding.

Quality control techniques

Dimensions

Components and products need to be checked to ensure that they meet their specification. The specification will usually relate to a special drawing of the product. This drawing is produced in a particular format or style that engineers are familiar with and understand, and is called an engineering drawing.

Components need to be checked to make sure they are the correct size. The size of each part of the component is written on the engineering drawing in the form of **dimensions**. Here we see an example of some dimensions.

Projection lines are drawn to mark the edges of the component and to show where each dimension extends to. They are drawn from the edge of the component (but not touching it) to the **dimension line**. The dimension lines run parallel to the edge being measured and extend to the projection lines on each side of it. Arrowheads show clearly where each dimension starts and finishes.

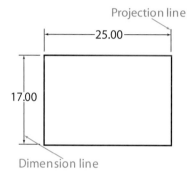

The dimensions do not have units such as millimetres (mm) or metres (m) after the number. This would make the drawing too cluttered. Instead of putting the units here, there is a small note on the drawing stating which dimensions are used. It is important that all the measurements on a drawing are in the same unit, e.g. 'All dimensions in mm'.

Dimensions can also be used to measure diameters of circles, angles or radii.

Examples of dimensions

Dimension lines and projection lines should be shown as clearly as possible on drawings. If there is any confusion an engineer or anyone else reading the drawing could mis-read it. Some rules are laid down which all drawings should follow to prevent any confusion.

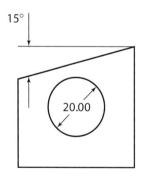

Examples of different types of dimensions

Here are some of the rules relating to projection lines and dimension lines.

- Projection lines must not touch the component drawing.
- Projection lines should not be crossed if possible.

- Projection lines should go slightly past the dimension line.
- The dimension should be placed in the middle of the dimension line.
- Dimensions should not be crossed by other dimension lines.
- Dimensions should be as clear as possible at all times.

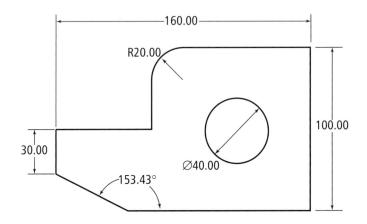

Tolerances

It is impossible to make components to the exact size shown on an engineering drawing. The component will always be slightly wrong, if only by a tiny amount. Engineers overcome this by adding a small amount either side of each dimension; anything falling within this range is then acceptable. This is known as a **tolerance**.

For example, if a product has a dimension of 120 mm, and it is made to 120.5 mm, it might be unacceptable. If the dimension is 120 mm but it can be plus 1 mm or minus 1 mm, then the manufacturer has a range of sizes that the product can be, in other words a tolerance, and 120.5 mm is acceptable.

This tolerance would be shown on a drawing in the format shown here.

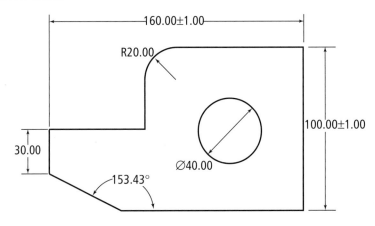

Here you can see that the dimension 160.00 has a tolerance of plus or minus (±)1.0.

Manufacturing tolerances

Each manufacturing process is capable of producing components to within a particular tolerance. When manufacturers are choosing which processes or machines to use to make their products, they need to consider how accurate they are. For example:

- precision machining: ±0.02 mm
- general machining: ±0.1 mm
- sheet metal work: ±1.0 mm
- drilling hole centres: ±0.5 mm
- sand casting: ±3 mm
- flame cutting: ±1.0 mm.

Complying with the fit

Components are assembled together to make a product. Each component fits to other components. Sometimes there is a slack fit and sometimes there is a tight fit.

You need to be familiar with the following terms relating to size when discussing fit:

- basic size (or nominal size): the reference size of hole/shaft to which the limits of size are fixed; the basic size is the same for both members of a fit
- limits of size: the maximum and minimum sizes permitted for a feature
  maximum limit of size – the greater of the two limits of size
  minimum limit of size – the smaller of the two limits of size.

The type of fit is extremely important. Types of fit are classified as follows.

Clearance fit

When two components are assembled, they can move freely. One component fits into a hole or cavity that is larger than the component. This could, for example, be for counterbores for screw heads.

Transition fit

When two components are fitted together there is no gap between the parts. This means that the components can move as they are not jammed together, but there is no free movement. This type of fit could be where components slide smoothly against each other such as machine slide-ways, or where components rotate, as in a bearing.

Interference fit

With an interference fit the two components are locked together. They are made so that not only is there no space between the parts but one component is bigger than the hole or cavity it is fitting into. This could apply to the point of a scriber that is forced into a scriber handle.

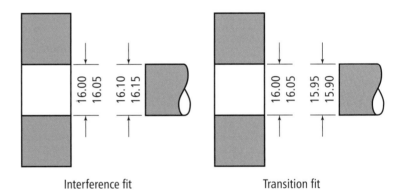

Example showing types of fits and their tolerance

Interference fit   Transition fit

To give the correct type of fit, it is important that the components are made to the appropriate tolerances. If both the shaft and the hole in the diagram above were exactly the same size, then the shaft would not slide.

In the example of transition fit, the smallest gap would be when the shaft is 15.95 mm diameter and the hole is 16.00 mm diameter, giving a total gap of 0.05 mm.

The biggest gap would be when the shaft is 15.90 mm diameter and the hole is 16.05 mm diameter, giving a total gap of 0.15 mm. This gap is not large enough to give a clearance fit, but is so small that there is a slide, or transition.

In the example of interference fit, there will be no gap as long the components are made to tolerance.

In the diagram at the top of the next page we see an example where the hole is 30 mm + 0.033 mm and the shaft is 30 mm − 0.092 mm to 30 mm − 0.040 mm. Here, the hole will always be bigger than the shaft if both parts are made to tolerance.

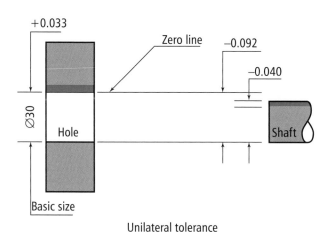

Unilateral tolerance

The basic hole system

Examples of commonly used fits for diameters from 6 mm to 180 mm

ISO fit	Code
Easy running	H7/e8
Normal running	H7/f7
Slide	H7/g6
Location	H7/h6
Push	H7/k6
Light press	H7/p6
Heavy press	H7/s6

The standard used for defining fits is BS4500. A shaft basis or a hole basis can be used. There is a special code for the type of fit required. With a hole basis, the code begins with 'H' followed by a letter and number.

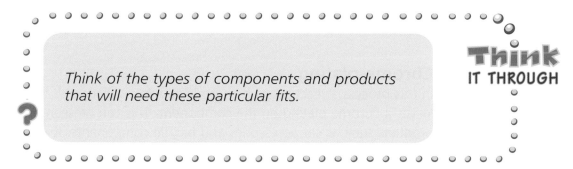

Think of the types of components and products that will need these particular fits.

Think IT THROUGH

Finish

When components have been manufactured the surfaces may be rough or smooth from the machining process. Surfaces may also have had some surface treatment or a coating of some type. The quality of the surface is known as the **finish**.

Coatings are designed to protect the component or to give a better looking finish – making something look good is known as **aesthetics**.

Cleaning and degreasing
The most basic operation to be carried out before a coating is applied to a component is to clean it.

Chemical degreasing
This removes chemicals and prepares the surfaces for other finishes.

Solvent degreasing

All contaminants on the surface of the component are dissolved in a solvent bath.

Vapour degreasing

Solvents are sprayed or vaporised onto the components to dissolve contaminants

Powder coating

A plastic (or polymer) powder is distributed over the surface of a component. Heat then melts the plastic to leave behind a high quality surface. The advantages are that the energy and labour costs are relatively low, it is environmentally safe and gives a good quality finish. Powder coating is often chosen in preference to painting as no solvents are involved.

The product specification should state the depth of coating, which can be tested.

Chrome plating

In this process a reaction with chromic acid takes place and leaves a chrome plating on the component. This can be seen on products such as car accessories and bicycle components. It gives a bright, shiny, silver coating which protects steel from corrosion.

Electroplating

Thin metal coatings are made by dissolving an anode in an electrolytic solution with an electric current. Several materials can be used as anodes. The material that is used will determine the coating on the surface of the component.

Precious metals such as gold and silver are used to coat jewellery.

Various metals can be electroplated onto the surface of a parent metal, including:

- copper (Cu)
- nickel (Ni)
- silver (Au).

There are some restrictions on the size of the surface that can be covered and the thickness of the coating that can be deposited.

Uniformity of thickness has a tolerance that is set by the process. This should be taken into account when setting tolerances on drawings for coating thickness.

Example of technical data with limitations of electroplating

Metal electroplating

Type	Adherence layers	Surface max	Thickness max	Uniformity*	Size of substrate
Cu	Ti/Cu, Cu	35 cm² in 100 mm 50 cm² in 150 mm	30 μm	±5% for 100 mm ±10% for 150 mm	100 mm and 150 mm
Ni	Cr/FeNi, FeNi, TiW, W, NiCr, Cr	20 cm²	20 μm	±15%	4″
Au	Electroplated Ni, Cu, Ta/Au, Cr/FeNi	20 cm²	4000 A	±10%	4″

*The uniformity strongly depends on design

Alloy electroplating

Type	Adherence layers	Surface max	Thickness max	Uniformity*	Size of substrate
FeNi 50/50	Cr/FeNi, Au	7 cm²	40 μm		100 mm
FeNi 20/80	Cr/FeNi, Au	10 cm²	15 μm	±10%	100 mm
SnPb 37/63	Electroplated Ni, electroplated Cu	20 cm²	15 μm	±15%	100 mm

Chemical symbols used in the table

- Au (gold)
- Cu (copper)
- Ta (tantalum)
- FeNi (iron/nickel)
- SnPb (tin/lead)
- Cr (chromium)
- Ni (nickel)
- Ti (titanium)
- W (tungsten)

Hot dip coating

With hot dip coating, components are dipped into molten metal which leaves a coating on the surface. Zinc is commonly used in this process and is a cheap and strong method of coating. It is often used for car bodies prior to painting, buckets, bins, lamp-posts and products designed for outdoor use.

Surface finishes can be described as **rough** or **smooth**. A surface finish is quite complex and can be described in terms of three qualities: **roughness**, **waviness** and **form**. It is often difficult to identify these qualities by looking at the surface of a component.

Roughness

This refers to surface irregularities resulting from the production process, rather than from the machine itself, such as marks produced by a cutting tool.

Waviness

This describes a component's texture. Roughness is superimposed on this. Waviness is due to characteristics of the machine such as the machine or work deflection, tool wear or vibration.

Form

This is the general shape of the surface, upon which the roughness and waviness characteristics are superimposed. Form errors are usually caused by the machine and the work piece flexing, or by slide-way errors.

Most surfaces are a combination of all three characteristics, but it is useful to measure each of them separately.

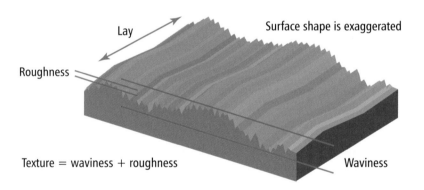

A magnified surface of a component

This diagram shows the relationship between the elements that make up the surface of a component. The pattern made by the texture is called the **lay**. The diagram above shows a unidirectional lay (the texture lies in straight lines). The lay can take various forms which will depend on the process used to produce the finish.

When surfaces are magnified many times they resemble a series of **peaks and valleys**, similar to a range of mountains. Surface roughness is measured as mean roughness or roughness average (R_a). This is also known as centre line average (CLA). This figure gives an average 'roughness' of all the peaks and valleys and is one way in which surface finish can be measured.

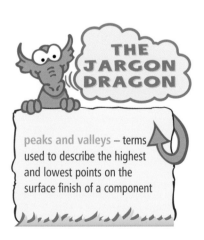

THE JARGON DRAGON

peaks and valleys – terms used to describe the highest and lowest points on the surface finish of a component

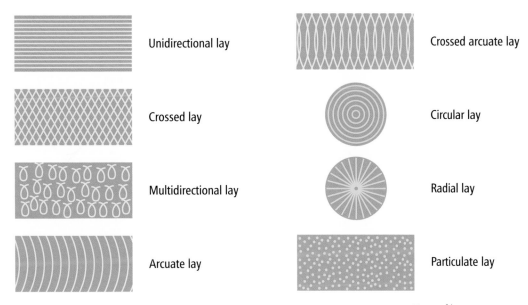

Types of lay

The units used to measure surface texture, such as the scratches on the surface of a component, are micrometres. A micrometre is one thousandth of a millimetre. Because these measurements are so small, specialised measuring devices are used. These can give an electronic readout or print out very quickly.

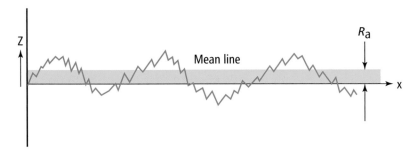

Surface finish symbols on drawings

When surface finish is measured against the engineering drawing, it must be clear on the drawing what the surface is intended to be. Standard symbols are used which give all the details about a surface.

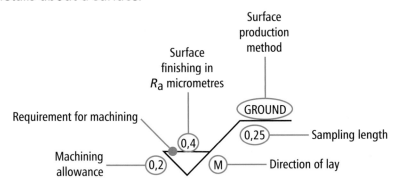

Machining symbols give roughness, lay and method of manufacturing to produce the surface.

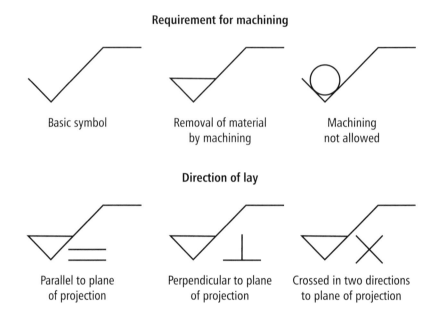

Requirement for machining

Basic symbol

Removal of material by machining

Machining not allowed

Direction of lay

Parallel to plane of projection

Perpendicular to plane of projection

Crossed in two directions to plane of projection

Micro-surface scales

Normally, a person can compare the texture of different surfaces by feel and sight. Designers and inspectors, on the other hand, use special scales to check product surfaces. A different scale is produced for each method of machining, and they can be used to directly compare surfaces.

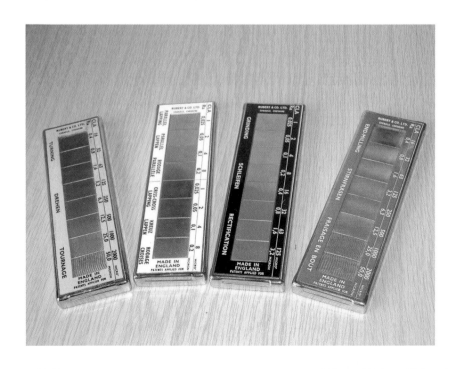

A range of micro-surface scales

Tools and equipment

It is important to be able to select the appropriate tools and equipment to undertake a task.

We can categorise these into three basic groups:

- marking-out tools
- hand tools
- inspection tools.

Marking out

If you were drawing on a piece of paper you would use pencils, erasers, compasses and set-squares. It is much more difficult to write or mark on to metal such as steel.

The metal to be worked upon is known as the **work piece**.

The process of drawing onto the work piece is known as **marking out**.

There are two main ways of marking out. In **flat marking out** the work piece is laid flat on a surface, and in **vertical marking out** the work piece is held in a vertical position.

THE JARGON DRAGON

marking out – drawing on metals using special ink and steel marking-out tools

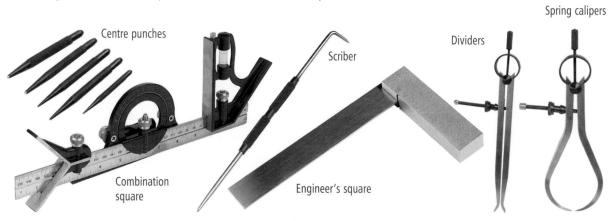

Centre punches

Scriber

Spring calipers

Dividers

Combination square

Engineer's square

A range of marking-out tools

Flat marking

This is the process for flat marking:

> If any surfaces that are supposed to be flat are damaged or there is dirt between the surfaces, then the marking out process will not be accurate. It is important therefore to protect and clean all the marking-out equipment

before it is used. The work piece must also be cleaned with a cloth to remove oil and grease.

De-burr the material. This means removing all sharp edges with a file.

Add **marking blue**. This is a dark-blue runny ink which is painted onto the surface of the metal. It dries very quickly. Once the marking blue is dry the work piece is ready to be marked.

A **scriber** is held like pen, and has a sharp steel point. It is used like a pen to mark lines onto the steel.

Lines drawn upon engineering products will be either straight or curved. A rule is used for straight lines. The rule is placed in position on the work piece and the scriber is used to mark the line.

When a line is needed to be at right-angles to the edge of the work piece, an **engineer's square** is used. The engineer's square is held in place by hand and the line is scribed with the scriber.

Often an angle is needed. In this case a **combination square** is used. This is used in a similar way to an engineer's square but it can be set to any angle.

Odd legs are used to produce a straight line parallel to the edge of the work piece. One edge has a step and the other has a sharp point. This sharp point does the work of a scriber and marks a fine line on the surface of the work piece.

Dividers are used in the same way as compasses are used on paper. There are two points, one on each end of the dividers. One point is placed in a small indented hole on the work piece. The dividers are then used to produce a circle or arc on the work piece.

The small indentation needed in order to use a pair of dividers is produced by a **centre punch**. A centre punch is similar to a scriber but much thicker and heavier. It is hit with a hammer to create a small indentation in the work piece. This indent may be used to locate the dividers in order to produce a circle or arc. A centre punch could also be used to produce a series of small indents along a scribed line. If the marking blue is worn from the surface of the work piece, the small indents produced by the centre punch can still be used as a guide.

The indentation produced by a centre punch could be used to locate a drill. This would require a bigger indent

A hammer and centre punch being used

than for the purposes described above. The drill bit is located in the indent and the hole is drilled in that position.

Vertical marking

Often it is better to stand the work piece up vertically in order to mark it out. This will require some special marking-out equipment.

A **surface table** is a perfectly flat table made of cast iron, steel or granite. Sometimes a smaller flat plate is used, known as a surface plate.

The surface table or surface plate must be protected from damage and is usually covered when not in use.

Vertical marking out must be done on either a surface table or a surface plate, as these are guaranteed to be flat, and not on a general workbench or desk.

Parallels are perfectly flat lengths of bar which sit upon the surface table. The work piece sits on top of these, which makes the process of marking out much easier.

An **angle plate** is an L-shaped device in which the angle is exactly 90°. The work piece rests against the angle plate, which ensures that it is perfectly vertical. The work piece can be either held against the angle plate or alternatively clamped in place using slots in the angle plate.

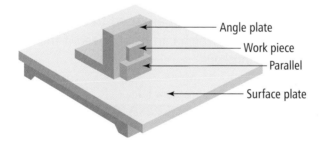

Angle plate
Work piece
Parallel
Surface plate

A **scribing block** holds a scriber in place. It has a flat base so it can sit perfectly flat on the surface table. When the **surface gauge** moves left to right the point of the scriber always remains at exactly the same height.

A **Vee block** is used to mark out a round bar. The round bar sits in the Vee block and is sometimes held in place by a small clamp.

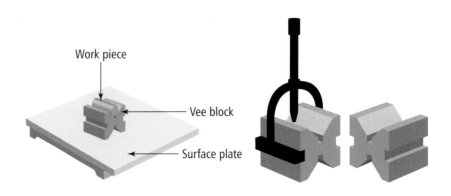

Work piece

Vee block

Surface plate

Hand tools

When marking out is complete, the component or product can be manufactured. This may involve using **hand tools** or machines.

Using hand tools requires skill, and this is gained through practice and training.

THE JARGON DRAGON

hand tools – the general tools in the workshop that do not need power

A selection of hand tools

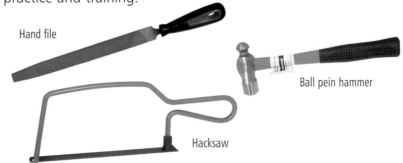

Hand file

Ball pein hammer

Hacksaw

Hacksaws

Hacksaws are special saws that are used to cut metals. They have a long thin blade which can be replaced if the blade becomes worn or broken.

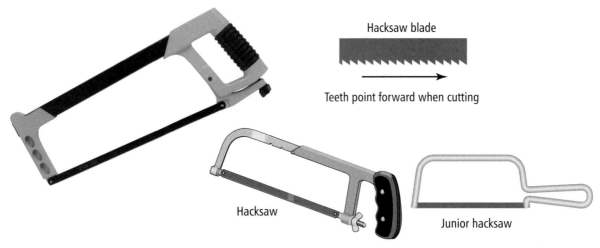

Hacksaw blade

Teeth point forward when cutting

Hacksaw

Junior hacksaw

Using a hacksaw

Here we see a hacksaw being used. This type is called a pistol-grip hacksaw, because it is held like a pistol. The blade points forward and is parallel to the ground. The hacksaw is moved backwards and forwards, though the teeth are designed so that the saw cuts the metal only on the way forward, away from the person holding the saw.

To move the hacksaw, one hand holds the handle and moves the saw backwards and forwards smoothly, like a machine. The other hand sits over the frame, without gripping it, and steadies the saw. This keeps the hacksaw pointing forward but does not apply a large downward force, as this may cause the blade to snap.

The number of teeth on a hacksaw can vary. Generally, the harder the material, the more teeth are needed to cut it. At least three teeth should be in contact with the metal to be cut at any one time, otherwise the hacksaw will 'bounce' across the material and it will be difficult to cut.

When using a hacksaw, at least three teeth must be over the work piece at any one time

Hacksaw blade   Hacksaw blade

Work piece   Work piece

Files

When metal has been cut with a hacksaw it will not have a very smooth finish. Files are used to produce a smoother finish by removing small amounts of metal down to the scribed line. This is much more accurate than trying to saw to the scribed line with a hacksaw.

Although files produce a flat finish, the quality of the finish depends on the skill of the user; it takes a lot of practice to be able to use a file accurately.

Files can be categorised in a number of ways:

- length of file
- type of cut
- grade of cut
- shape of file.

Length

The length of a file is measured from the point to the shoulder and does not include the handle.

Type of cut

Files can be **single cut**, with one set of diagonal teeth, or **second cut** with two sets of diagonal teeth opposite each other.

- **Rasp files:** these are generally used for soft materials such as wood.
- **Needle files:** small, accurate files for precision work.
- **Swiss files:** small files for accurate work.

Grade of cut

The grade of the cut is based upon how many teeth are in a 25 mm length of file.

Grade of cut	No. of teeth per 25 mm
Rough	12–20
Bastard	14–40
Second cut	25–52
Smooth	35–62
Dead smooth	60–88
Superfine	100–170

Shape of file

There is a wide range of different files that are used to cut different shapes.

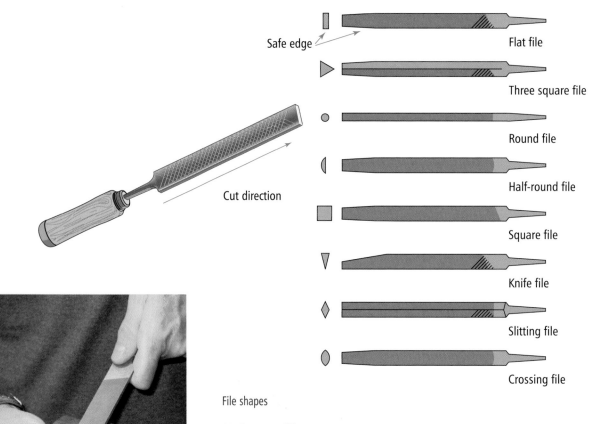

File shapes

Using a file

The user stands facing the work piece and firmly holds the file handle in one hand, resting the other hand on top of the file. As with the hacksaw, the file is moved backwards and forwards. If a flat finish is required then the file must always be horizontal and parallel to the ground.

Unlike a hacksaw, a file can be used in different directions. Cross filing helps metal removal and prevents vibration. Draw filing gives a better surface finish.

Hammers

Hammers come in a large variety of shapes and sizes. There are several types that have specific tasks.

Ball pein hammers

These are widely used. The flat end is used for **chiselling** or to hit a centre punch. The **ball pein** (the domed end) is used to hammer rivets.

Hammer surfaces are made to be very hard. If a hammer hits another hard steel such as another hammer surface, it can shatter or fracture, so care must be taken when using them.

Hammers come in various weights and sizes, and it is important to use a hammer that is the correct weight. You should consider how the weight will affect the work piece and how comfortable you are with handling the hammer.

Mallets are used with soft surfaces to deliver a powerful blow but not damage the material they are hitting. Mallets are made of materials like hide, copper or nylon. They are used to knock components together that have an interference fit, such as bearings and gears on shafts and keys in keyways. A copper mallet is used for keyways to force the components together. Copper is a softer metal than steel so it will dent slightly instead of spoiling the component it is hitting.

When bending hot metal, a hide mallet is heavy but soft enough not to fracture the hot metal. The heat is easily dissipated through the hide.

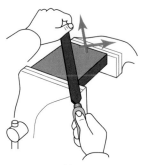

Cross filing

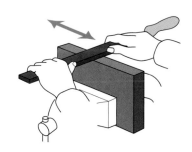

Draw filing

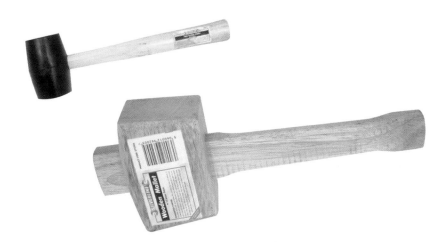

Chisels

Chisels are used to remove metal. They are placed in the correct position required and hit on the 'head'. The point of the chisel pierces the work piece and slices off a metal chip. These chips are extremely sharp and can fly very quickly. Goggles must **always** be worn when chiselling.

Chisels are not very accurate but have the advantage of removing relatively large amounts of metal. They can also be used to cut or 'shear' through components, such as bolts, which cannot be loosened.

The angle at which the chisel enters the work piece depends upon the type of metal that the work piece is made from. This angle is known as the **angle of inclination**. The angle of the point is know as the **cutting angle** and also varies with the material being used.

Material	Cutting angle	Angle of inclination
Cast iron	60.0°	37.0°
Mild steel	55.0°	34.5°
Medium carbon steel	65.0°	39.5°
Brass	50.0°	32.0°
Copper	45.0°	29.5°
Aluminium	30.0°	22.0°

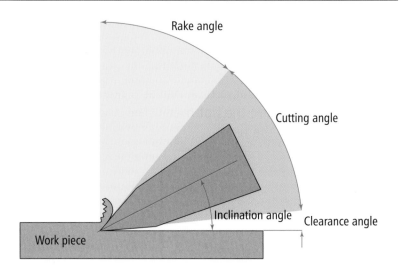

Important angles when chiselling

Chiselling

Flat chisels are for general use. **Cross-cut chisels** are designed to cut deep grooves for keyways.

The head of a chisel gradually becomes mushroom shaped with use. There is then a danger that it may fracture and a small part could fly off and cause damage. It is important therefore to always make sure that the chisel head is **chamfered** and not mushroomed before using it.

Remember:

- Always cut with the chisel moving away from you.
- Ensure that no one is in front of the work piece.
- Ensure that the chisel head is chamfered and not mushroomed.
- Always wear goggles.

Centre lathes

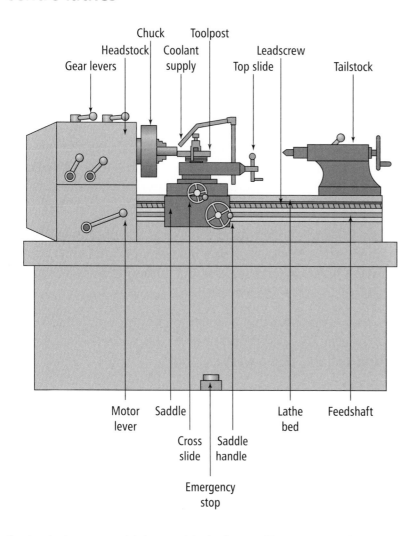

Centre lathes are widely used in industry. They are used to produce cylindrical shapes. The work piece revolves around its own centre. A **tailstock** sits at the back of the **lathe bed**. **Chucks** and **drills** can be put into the tailstock. The centre of the drills are at the same height as the centre of the work piece.

Tools that cut the work piece are also at the same height or 'on centre' The fact that everything revolves around the centre of the work piece gives the centre lathe its name.

Centre lathes are very accurate, but as with all equipment their accuracy depends on how well the machines are maintained and the quality of the cutting tools used. If the cutting tool is worn then the accuracy of the lathe will be reduced.

The work piece is held in a chuck specially designed to hold round bars. The chuck revolves and the cutting tool is moved in to a position where it will cut the revolving work piece. The process of using a lathe is known as 'turning'. A person who is specially skilled in the use of centre lathes is known as a 'turner'.

There are several important things that you need to consider when using a centre lathe.

Type of turning tool

The **turning tool** is held in the toolpost. The shape of the turning tool will determine the type of finish and the form of the work piece. It will also determine how much material can be removed.

Turning tools can be left-hand tools or right-hand tools. These cut in opposite directions. If you hold a right-hand knife tool in your right hand you can see that the cutting edge will cut as the tool moves towards the left.

Tools may need to be thin enough to cut grooves in the work piece, or even to cut right through it. This process is known as **parting off** and uses a **parting tool**.

Types of turning tool

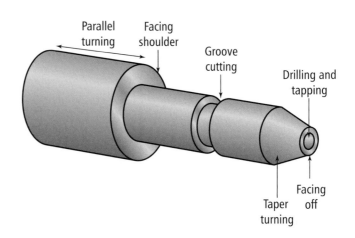

Types of turning operations

Operations

Many operations can be performed on a centre lathe, and these can be categorised into a few very basic operations.

- **Parallel turning:** this procedure removes material from the work piece so that it has a smaller diameter. The diameter will be the same all the way along its length.
- **Facing off:** when the work piece has been sawn it will have a rough edge. Facing off moves the tool across the front face of the work piece making it 90° to the length of the shaft. This process can be used to make sure that all faces and shoulders are at 90° to the shaft.
- **Taper turning:** when a work piece needs to be a smaller diameter at one end than the other, a **taper** can be turned. For short tapers the cross slide is moved to the angle required. As the cross slide hand wheel is revolved the cutting tool moves along the work piece and cuts a taper.

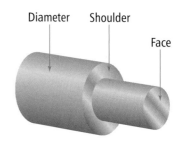

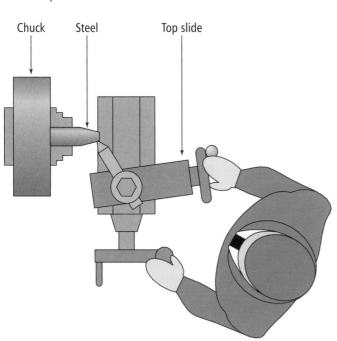

Drilling

A centre lathe can also be used to drill holes. The drill is placed into a small drill chuck in the tailstock, which sits on the back of the lathe bed. The tailstock is moved nearer to the work piece by unlocking the quick-release lever and holding it firmly in place. A hand wheel is then rotated which moves the drill bit into the work piece.

The depth of hole is read from a scale on the tail stock.

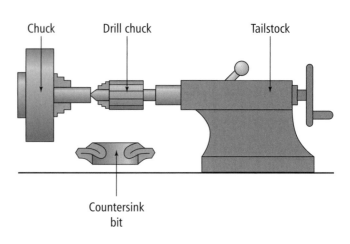

Chuck   Drill chuck   Tailstock

Countersink
bit

Knurling

When a grip finish is needed for products such as handles or hammer shafts, a process known as **knurling** can be used.

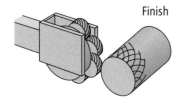

Finish

A knurling tool has two rollers with teeth formed in the surface of the rollers which are forced onto the surface of the work piece as it revolves. The result is a grip texture which has a pattern of small diamond shapes cut into the work piece.

Milling machines

Milling machines are generally large machines and are used widely in industry. The work piece is held in a machine vice which sits on a work table. The work table is moved left to right and forwards and backwards. The work table can also move upwards.

A cutting tool revolves but stays in the same position. As the work piece moves past the cutting tool, material is cut from the work piece.

Milling machines are very accurate but their accuracy depends upon the quality of the cutting tool and the care that is taken to maintain and clean the machines.

Milling machines can be vertical or horizontal depending upon whether the spindle and the cutting tool are in a vertical or horizontal position.

Vertical milling machine

The vertical milling machine is used to cut slots, square edges, shoulders and holes.

The cutting tool comes in a variety of different forms designed to produce different cutting profiles.

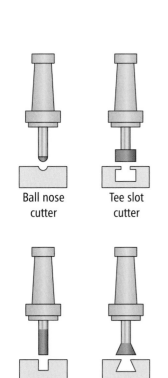

Ball nose
cutter

Tee slot
cutter

Slot drill

Dovetail
cutter

A range of vertical milling cutters

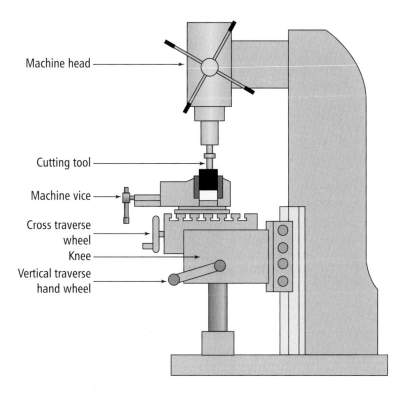

Machine head

Cutting tool

Machine vice

Cross traverse wheel

Knee

Vertical traverse hand wheel

A vertical milling machine

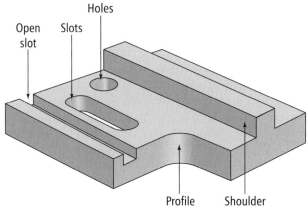

Holes

Open slot

Slots

Profile

Shoulder

Typical vertical milling operations

Horizontal milling machines

Horizontal milling machines are less widely used than vertical milling machines but are very effective at producing slots and keyways.

In a horizontal milling machine the cutting tool rotates on a horizontal spindle known as an **arbor**. Cutting tools can be grouped together on the arbor, which is a very efficient way to produce work with a continuous profile or cross-section.

Horizontal milling cutters can create **chamfers** and **radii** on the edges of work pieces. They can also produce flat-bottomed, v-shaped or radiused grooves.

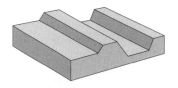

Typical horizontal milling operations

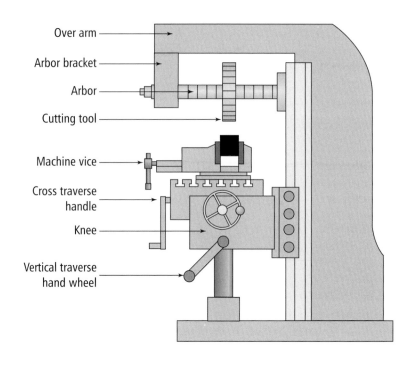

Over arm

Arbor bracket

Arbor

Cutting tool

Machine vice

Cross traverse handle

Knee

Vertical traverse hand wheel

A horizontal milling machine

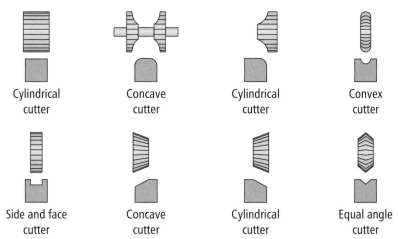

| Cylindrical cutter | Concave cutter | Cylindrical cutter | Convex cutter |

| Side and face cutter | Concave cutter | Cylindrical cutter | Equal angle cutter |

A range of horizontal milling cutters

Drilling machines

Drilling machines are extremely common in engineering workshops. Their role is to produce holes in work pieces. There is no accurate way to move the work piece once it is clamped so the marking out of the work piece is extremely important.

There are two types of machine drill: bench drills and pillar drills.

Bench drills

Bench drills are secured to the top surface of the work bench or a specially produced frame. There is limited adjustment on the height of the table. They have a range of speeds so that different-size holes can be drilled. They are smaller than pillar

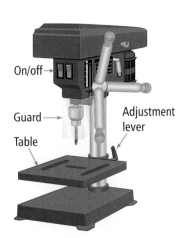

On/off

Guard

Table

Adjustment lever

drills and are generally less expensive, which means that they are often used in workshops at home.

Pillar drills

Pillar drills operate in the same way as bench drills, but the table can be adjusted down almost to the floor, as it is attached to a pillar. This allows larger or taller work pieces to be drilled. Generally, pillar drills are much more robust, rigid and powerful than bench drills and can therefore drill bigger holes.

Drilling tools

Drilling machines use **twist drills**. These come in standard sizes from 0.5 mm up to 13 mm to be used in a drill chuck. Twist drills that are bigger than this in diameter are held directly in the spindle of the drill by a **morse taper**. Twist drills held in place by a morse taper are released using a drill drift and mallet.

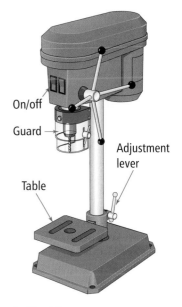

On/off

Guard

Adjustment lever

Table

A pillar drill

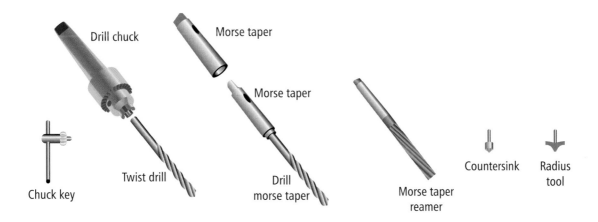

Drill chuck

Morse taper

Morse taper

Chuck key

Twist drill

Drill morse taper

Morse taper reamer

Countersink

Radius tool

Drilling tools

Countersink

A countersink produces a chamfer leading in to a hole that has been drilled. This allows countersunk screw heads to lie flush with the surface of the work piece. The size of the chamfer depends on how deep the countersink is pushed into the hole.

Counter bore

A counter bore drills a second hole over the drilled hole at a set depth. This allows screw heads to sit below the surface of the work piece.

Spot face

This is where a slot drill cuts a round, flat surface onto the work piece. This allows fasteners such as washers or screw heads to be tightened onto a flat surface.

Centre drills

Centre drills are used to give a lead in to bigger drills. When a small hole is required, a centre punch indent is enough to align the drill. When a larger hole is needed, a centre drill creates a guide hole big enough so that the larger drill will follow.

Work-holding devices

When work pieces are to be machined they need to be held securely in place.

Here are some questions that will help you decide which device to use for a particular job:

Where is the work piece to be held?

- on a work bench
- on a centre lathe
- on a milling machine
- on a bench or pillar drill?

What is the shape of the work piece?

- round or cylindrical
- square or flat
- odd shape?

Bench vice

A **bench vice** is generally used on work benches in the workshop. They are strong and sturdy so can withstand the wear and tear of the workshop. They are very heavy, but they are rarely moved from place to place. They are not very accurate but are perfect for hacksawing and filing.

Bench vice

Holding work pieces on a bench drill or pedestal drill

A hand-held vice can be used if the holes to be drilled are very small. They are designed to withstand the downward pressure of a drill as the work piece rests on a surface. There are usually slots in the base where clamps can be located to hold the vice if necessary, for example when the holes to be drilled are quite large.

Hand-held vice

Milling vices

Vices for milling machines always need to be clamped to the work table as milling machines are very powerful. The jaws that hold the work piece are flat and accurate. A handle is needed to tighten the work piece. There are two main types of vice: a fixed type or a universal type. Universal milling vices can tilt to an angle or rotate.

Fixed milling vice

Universal milling vice

Machine clamps

For awkward or irregular-shaped work pieces, clamps can be used. These come in many forms to allow for as many shapes of work piece as possible. They sit onto the work piece and the work piece is held in place with bolts.

A range of machine clamps

Angle plates

Work pieces may need to be clamped at 90° to the work table. For this an angle plate is used. Slots in the angle plate allow the work piece to be clamped.

An angle plate

Chuck key

Work holding on centre lathes

Special devices are used to hold bars in a centre lathe. Round bars are held in a **3-jaw chuck** and tightened with a **chuck key**. The 3-jaw chuck always locates the work piece on the centre line of the centre lathe.

A **4-jaw chuck** is used when the bar needs to be off centre in order to produce components with eccentric parts.

3-jaw chuck

4-jaw chuck

Centre lathe face plate

Centre lathe face plate

If an odd-shaped work piece is to be machined on a centre lathe then a **face plate** is used. The work piece is clamped to the face plate. This system is used when drilling, boring or turning sections of an odd-shaped work piece.

Computer-aided machining

Modern manufacturing demands better **productivity**. Using computers, products can be designed and then programmed to suit a variety of special computerised machines.

The speed and accuracy of machines such as milling machines and centre lathes can be improved if they are controlled by computers and software. This system is known as **computer numerical control (CNC)**. Computer-controlled machines work in a similar way to conventional machines, but the various functions they perform are done automatically in response to a program. For example, they move to new positions automatically; spindles start and stop automatically; and coolant is switched on and off automatically. Many CNC machines also have automatic tool change facilities and automatic tool measuring devices.

The complexity of the automation of CNC machines means that they need to be powered by electricity, pneumatics and hydraulics. As with conventional machines, maintenance is extremely important.

THE JARGON DRAGON

productivity – the amount produced, or work done, for a given cost

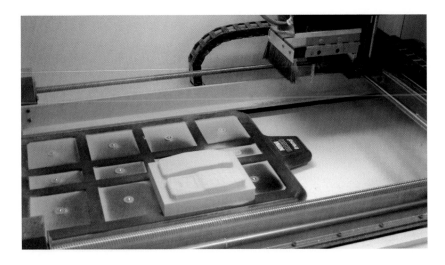

A work piece is placed in the machine vice of a CNC machine centre

Modern CNC machines have an input terminal where programs can be input manually or by a direct line from a **computer-aided design (CAD) system**. The programs that control the CNC machines are extremely complicated and it takes a highly trained programmer to produce them.

Because computer-controlled machines operate automatically, they generally have much better machine guards than conventional machines, which protect operators from moving parts.

CNC machines are extremely useful where large numbers of components are to be manufactured or where complicated shapes require machining. However, CNC machines and the tools that they use are very expensive, and so engineers have to balance the financial cost of buying a CNC machine against the benefits they offer in terms of production.

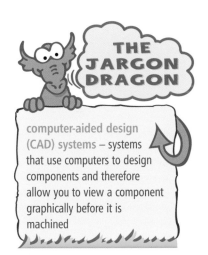

THE JARGON DRAGON

computer-aided design (CAD) systems – systems that use computers to design components and therefore allow you to view a component graphically before it is machined

Computer-integrated manufacture
CNC machines can be used in conjunction with robots to create automatic production systems known as **cells**.

Although automation reduces the need for large numbers of operators, computer-aided manufacturing involves a number of highly skilled setters and maintenance personnel.

Comparison of tool machinery costs

	Hand tools	Machines	CNC machines
Cost	Low	Moderate	High
Accuracy	Low	Good	Excellent
Skill to set	n/a	High	High
Skill to operate	High	High	Low

A modern CNC machine tool showing tools mounted in a carousel

Measuring tools

When components have been produced they will need to be measured. There is a wide range of measuring tools available for measuring lengths, diameters, depth and angles. It is important that you are aware of what tools you can use for each type of measurement.

Measuring tools are usually described in terms of their **accuracy**, **range** and **repeatability**, as well as **expense**, **availability**, **ease of use**, **storage** and **calibration**.

300 mm rule

Digital caliper

Vernier caliper

Tape measure

A selection of measuring tools

Generally, the more accurate a measuring tool, the more expensive and difficult it is to use.

Measuring length
Steel rules
Steel rules for use in workshops come in a range of sizes, from 150 mm to 1 m long. They are accurate to 0.5 mm.

Steel rules are specially designed for measuring components – they have a square end, which means that they can be pushed up to the shoulder of a component or lined up with the end, and the scale starts from the very edge of the square end. Rules that are used for making measurements on diagrams are not suitable for measuring engineered products as they do not have a square end. Steel rules are very good for measuring length but are used only for very approximate measurements of diameter as it is difficult to know if the rule is on the centre of the bar.

Steel rules are relatively cheap and easy to use but are not very accurate. With machining, they are used only for approximate measurements. When products are being fabricated or formed using presses, mallets and hammers, though, measurements do not need to be so precise, so a steel rule is suitable.

Vernier calipers

Vernier calipers are used where more accurate measurement is required. Vernier calipers for general workshop use range in size from 150 mm long to 600 mm long. Their accuracy depends on the skill and eyesight of the user, but generally they are accurate to 0.1 mm.

Calipers are also available with dial readouts and digital readouts, although this adds to the price. Calipers are very versatile and for this reason are widely used. They can measure lengths, outside diameters, inside diameters, depths of holes, slots and steps.

Measuring diameters

External firm joint calipers are designed to open up easily to fit over a diameter such as the end of a bar. The calipers are pushed over the bar until the jaws slide smoothly with no gap. There is no scale on the calipers so they must be used with a steel rule to give a measurement.

Internal firm joint calipers are designed to measure holes. They are inserted into the hole, opened up to fit the hole, and the measurements read with a steel rule.

Spring joint calipers can be internal or external and are used in the same way as firm joint calipers. However, they also have an adjusting screw which allows finer adjustment of the legs. This makes the calipers easier to fit than firm joint calipers. Spring joint calipers are read by using a steel rule to measure the gap. This gives the same accuracy as firm joint calipers.

Vernier caliper

Digital caliper

Micrometers

Micrometers come in a variety of sizes. Each type can measure a range of only 25 mm. Different types of micrometer are named according to the range of measurements that they can measure, for example a '25 to 50' micrometer measures the range 25 mm to 50 mm.

Micrometer sizes include:

0–25 mm
25 mm–50 mm
50 mm–75 mm
75 mm–100 mm
100 mm–125 mm
125 mm–150 mm

These are the general workshop sizes, although micrometers can be much bigger.

A micrometer has a scale along the main diameter known as the **barrel**. This gives the length to the nearest half millimetre. As the **spindle** rotates another scale moves. When the spindle stops on the component a reading is taken.

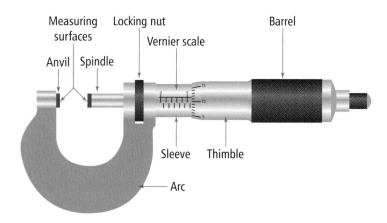

Measuring surfaces — Locking nut — Vernier scale — Barrel
Anvil — Spindle — Sleeve — Thimble — Arc

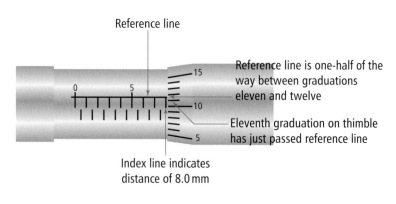

Reference line

Reference line is one-half of the way between graduations eleven and twelve

Eleventh graduation on thimble has just passed reference line

Index line indicates distance of 8.0 mm

How to take a reading from a micrometer

Health and safety

Throughout this unit you will have seen that health and safety are extremely important. For each topic you've learned about in this unit, you will have been given the relevant health and safety information. This section gives more general information about health and safety issues.

Working safely

You must always make sure that you are working safely and that you are not endangering yourself or others. This includes the other students in the workshop, teachers and technicians.

Behaviour

There are many dangers you need to be aware of when working in a workshop situation, and you need to behave responsibly and maturely at all times.

Be aware of:

- sharp tools and objects
- moving machinery such as lathes, milling machines and drills
- slippery floors from oil or coolant
- hot material from a forge.

Remember that behaving irresponsibly in these situations can lead to serious injury or even death.

Workshops are noisy and full of moving machinery. They can become very hot if a forge is being used, or they can be very cold through the fact that they are usually in large open spaces. This can make you feel uncomfortable or tired, or lose concentration. Be aware of this. Tell someone if you are feeling tired or unwell.

Around most workshops you will see lines on the floor. These are the walkways. Always stay within the walkways unless you need to use machines that are off the walkways.

Emergency stop buttons

These are found all around workshops. Pressing an emergency stop button will turn off the power to every machine in the workshop. They are therefore only used in an emergency. They can sometimes be knocked by mistake, which can cause problems for anyone using a machine.

Think IT THROUGH

Look for all the emergency stop buttons in your workshop. Show them on a plan. How many are there?

There are various types of emergency stop button, but they always have a large red button to press. Some have a key so that only the person with the key, usually a supervisor or teacher, can switch the power back on.

If you see someone in danger you should hit the emergency stop button and tell the supervisor or teacher immediately.

Switching off an emergency stop button is a serious matter – never do it unless you are sure it is really necessary.

Supervision

Students should always be supervised in the workshop. If there is no teacher present, they should not enter the workshop.

Students should be careful when using hand tools with sharp edges such as scribers, chisels and hacksaws. Before using machines, students must inform a teacher, as some machines need constant supervision. The teacher will have made a **risk assessment** of all the machines in the workshop, so they will know how much supervision is required.

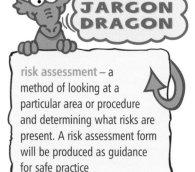

THE JARGON DRAGON

risk assessment – a method of looking at a particular area or procedure and determining what risks are present. A risk assessment form will be produced as guidance for safe practice

Personnel

There are many people who could be working in the workshop at the same time as you.

- **The teacher:** this is your main supervisor at school. During college activities they will work with a college lecturer.

- **College lecturer:** this is your main supervisor when you are at college.
- **Technician:** their role is to support the teaching staff. They perform many tasks, including cutting material and preparing machines.
- **Stores person:** their role is to look after all tools in the tool store and give out materials, tools and equipment to students and support staff.
- **First aider:** this person treats any injuries that occur.
- **Inspectors:** their role is to check that teaching staff are teaching what they are supposed to be teaching and that students are learning.
- **Other students:** there may be other student groups in the workshop. Some will be older and some younger. Respect all other students using the workshop at the same time as you

Personal protective equipment

This term applies to any equipment that is used to protect you.

Anyone working in a workshop should have safety boots and overalls. Goggles should always be worn when using machinery or when there is a danger of sharp or hot objects hitting the face.

Safety boots, overalls, goggles, helmets, ear defenders and gloves – just some of the safety equipment designed to prevent you from being injured

Working with heat

When using heat, for example when you are heat treating metals, you will need to wear special leather gloves that protect your hands and arms.

If you are brazing or welding you will need special heat protection in the form of a welding mask and gloves.

Fire safety

Different types of fire extinguisher are made for different purposes. It is important to recognise what sort of fire each type is used on.

Here we see a range of fire extinguishers.

Water	**Powder**	**Foam**	**Carbon Dioxide (CO₂)**
For wood, paper and solid material fires DO NOT USE on liquid, electrical or metal fires	For liquid and electrical fires DO NOT USE on metal fires	For use on liquid fires DO NOT USE on electrical or material fires	For liquid and electrical fires DO NOT USE on metal fires

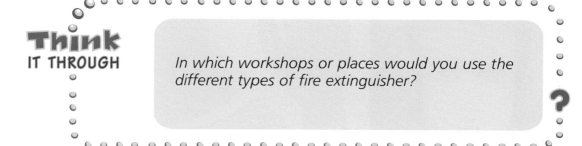

Think IT THROUGH

In which workshops or places would you use the different types of fire extinguisher?

Risk assessment

You should be able to look around a workshop and identify potential hazards so that you can take precautions to prevent accidents happening. This is known as **risk assessment**. There are five steps in risk assessment.

Step 1: Look for hazards
There are many places where accidents could occur. Look around the workshop. Can you spot anything that might cause an accident?

Step 2: Decide who might be harmed and how
Think of all the people who could be in a particular workshop or area. How could they be hurt? Often, if people are unfamiliar with a place, they can be more at risk than those who know it well.

Step 3: Evaluate the risks and decide whether existing precautions are adequate or whether more should be done
There may well be precautions in place, but ask yourself whether these would be enough and whether the procedures could be improved.

Step 4: Record your findings
Write your findings down so that you can refer to them at a later date.

Step 5: Review your assessment and revise it if necessary
As time goes by, look at your risk assessments again. Has anything changed? Do the procedures need to be changed?

Health and safety procedures

You need to be able to follow health and safety procedures and instructions. Sometimes instructions can be complicated. Don't worry, remember:

1 Listen.
2 If you don't understand instructions ask the teacher again.
3 If you cannot complete a task, stop and ask again.

Think safety all of the time!

You will be given instructions on how to complete tasks. The tasks will become more complicated the more you learn. You will learn easy procedures first, such as marking out and using hand tools, and when you can understand these you will be instructed on how to use machines such as drilling machines. You will then move on to automatic machine such as milling machines and lathes.

There are various safety procedures that must be followed.

Fire drills

Fire drills are used to practise getting out of a building in the event of a fire. You will be given instructions about what to do. Fire instructions are usually displayed on the walls in workshops. If you cannot find them ask a teacher.

Remember:

- When the bell rings leave the building with the teacher.
- Do not collect your belongings.
- Assemble at the assembly point given in the fire instructions.
- Your teacher will call the register.
- Do not re-enter the building until told to do so by your teacher.

If you spot a fire sound the alarm.

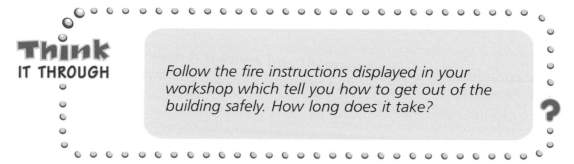

Think
IT THROUGH

Follow the fire instructions displayed in your workshop which tell you how to get out of the building safely. How long does it take?

House keeping

House keeping refers to all those activities that keep your work area clean and in good order. Good house keeping leads to a safer working area and more accurate work.

House keeping can include:

- keeping your workshop area clean
- making sure that boots and overalls are not left on the floor
- making sure that coats and bags are kept in lockers
- cleaning up spillages
- clearing walkways
- storing tools properly in the store room
- cleaning and oiling machines after use
- keeping washing areas clean.

Maintenance of tools and equipment

Engineering tools are expensive, and the better their quality, the better the work they can produce. All tools should be maintained which means being regularly cleaned, lubricated and stored correctly.

Different types of tools and equipment need to be maintained in different ways.

Marking-out equipment

Marking-out equipment is used to mark fine points and straight lines. It is therefore important to look after it. Scribers and centre punches should be placed in their individual holders after use. Scribers have fine points that are easily damaged, and their points should be checked periodically for sharpness. If they are not sharp they should be ground. This should always be done under supervision.

Steel rules are made from thin steel which means that they are easily bent and dented. They are often hung up by a small hole at one end, which prevents them being damaged by other tools. To protect steel rules, avoid laying them over other tools, and clean them with a cloth after use to remove oil and dirt.

Engineer's squares and rules need to be protected against damage to their edges. They should be stored in specially allocated areas, never on top of each other.

Hand tools

Hacksaws have fine blades which can easily be broken if they are not stored properly. Before using them, you should check that the teeth are not damaged.

Hacksaws are usually hung up on hooks to prevent other tools laying on their blades.

Files are extremely hard but also very brittle. The can easily chip, crack or break. They should never come into contact with other files as this could damage them. They should be cleaned with a **file card** after use and stored in a **file stand**.

THE **JARGON DRAGON**

file card – a flat wire brush used to clean hand files

Some sets of tool are made up of many pieces, such as socket sets, screwdriver sets and electrical installation sets. These are kept in special boxes where each part has its own place, so it is easier to check that all the parts are present. Tool sets should be checked before and after use to make sure no parts are missing.

Keeping your tools in a toolbox like this one helps you to find the tool you want, and to spot if anything is missing

Measuring equipment

It is very important to maintain measuring equipment, as people need to be able to rely on its accuracy.

Precision instruments such as micrometers, Vernier calipers and gauges are kept in individual boxes. They should always be cleaned with a cloth after use and returned to the stores. In industry, they are regularly checked against standards to make sure they are still accurate.

Machinery

Machines use oils and coolant. If left on a machine, these can cause damage and create problems. You should expect a machine to be clean when you start work, and you should always clean the machine after use.

Cleaning machines

- Always isolate the machine before cleaning This means switching the electrical supply to the machine off by an isolator switch on the wall.
- Remove the cutting tool – there is a danger of cutting yourself on the sharp edges of a tool as you clean the machine, so always remove this first.
- Clean the work table and machine vice (or clean the chuck on centre lathes). Use a hand brush and a cloth to remove all **swarf**, dirt and coolant.
- Clean all the slide-ways.

THE JARGON DRAGON

swarf – the waste material that comes from machined material. It is usually very sharp and can be hot

- When all oil and swarf have been removed, wipe a thin film of clean oil along the slide-ways.
- Empty the tray – remove all the swarf from the base of the machine and brush it all out into a bin.
- Clean the floor area.
- Remove the mat in front of the machine. Use a brush to clean the oil, swarf and coolant from the floor.
- Clean all the controls using a clean cloth.
- Clean the guard – guards should be clear so that the work piece can be seen.

What's in this unit?

To complete this unit you will need to understand the impact of new technology on engineering and manufactured products. You will learn that, whilst products can be categorised into different sectors of engineering and manufacture, they all utilise new technology.

You will investigate the impact of information and communications technology on businesses and you will investigate the impact of new components and a range of modern materials, including smart materials, in the engineering industry. You will learn about the changing systems of control technology and how modern engineering has developed because of this technology.

Application of Technology

In this unit you will learn about:

The manufacturing and engineering sectors

There is such an extensive range of products available that they are usually identified by grouping them into the following categories or **sectors**.

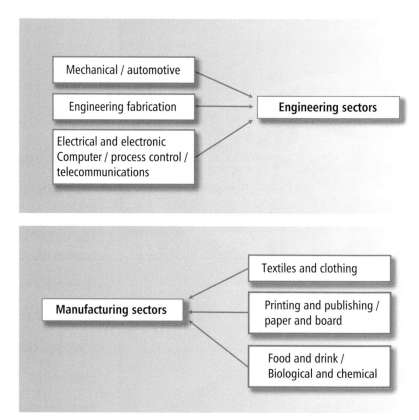

Engineering sectors

The **mechanical/automotive sector** includes products that are powered mechanically or automatically. Examples of products that fit into this sector include cars, vans, drills, presses, motorbikes, forklift trucks, power tools and many industrial machines.

The **fabrication sector** is perhaps the most difficult to classify. Products within this sector are generally defined as those that are made directly from raw materials such as sheet, plate, strip and bar. Generally, fabrication is associated with the joining and assembly of metal products, but other materials such as plastics and wood are also widely used.

Products in this sector include furniture, car parts, bridges, oil modules (rigs), hand tools, railings and gates, shopping trolleys, goal posts and mountain bikes.

Gas rigs are huge fabrications requiring vast amounts of labour, materials and equipment. They cost many millions of pounds to fabricate

A more familiar fabricated product is a mountain bike or BMX. These products are generally made of metal tubing formed and cut into shape before being joined together using a welding process

Products that contain electrical or electronic components such as TVs, DVDs, microwaves or other kitchen appliances, lighting and CD players fit into the third sector – **electronic and electrical, computer and telecommunications**. Games consoles, laptop, personal and palmtop computers are all examples of the computer element of the sector, while Internet modems, digital and mobile phones are examples of products in the telecommunications sector.

The sector of electronic and electrical, computing and telecommunications is one of the fastest developing in technology. Just think of a computer five years ago compared with the latest models – the computer above is small enough to hold in the palm of your hand!

Manufacturing sectors

The first of the manufacturing sectors is concerned with the production of fabrics, textiles and clothing. The **textiles and clothing sector** contains a huge number of different products, including all fashion such as clothing and footwear, as well as curtains, upholstery, carpets and many others.

Every time you read a book, a newspaper or a magazine, or write in a notebook or diary, you are using products from the **printing and publishing sector** of manufacturing. The many different types of paper and board used for packaging, toilet rolls and tissues are all examples of paper and board production.

The textiles and clothing sector of manufacturing rely heavily on the fashion industry for the promotion of their products

A few examples of the vast numbers of magazines that are included within the printing and publishing sector of manufacturing

The final sector is arguably the easiest of the six to classify – **food and drink, chemical and biological**. Popular products such as cola and lemonade, crisps, chocolate, in addition to other fresh, frozen and tinned food, all fall into this category. Examples of chemical and biological products would include toiletries such as shower gel and deodorant, household cleaning products, and pharmaceutical products such as headache tablets and cough medicine.

Various well known household products that fit into the sector of food and drink, chemical and biological

Each of the segments in the diagram below represents a sector of engineering/manufacturing. Look at the pictures of the products and decide which sector they each belong to.

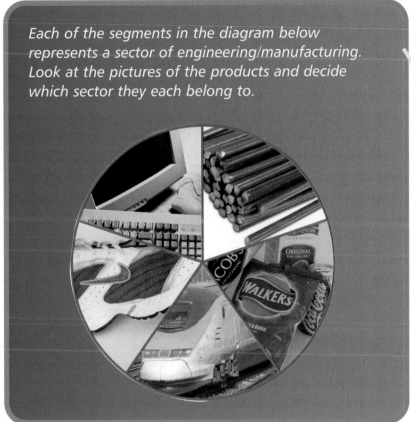

Turnover of the major sectors of engineering and manufacturing

The table below shows a breakdown of the major industries associated with engineering and manufacturing, and their annual **turnover**.

THE JARGON DRAGON

turnover – the total sales achieved by a company or organisation

Industry sector	Rank	Turnover (£ millions)
Food products, beverages and tobacco	1	75,019
Electrical equipment	2	65,007
Transport equipment	3	60,874
Chemical products	4	47,530
Printing and publishing, paper and paper products	5	45,981
Metals and fabricated products	6	41,081
Manufacture of machinery and equipment	7	33,522
Petroleum products and nuclear fuel	8	26,383
Polymer production	9	19,673
Textile production	10	13,502

The 'rank' column shows the position of each sector when they are ordered according to gross turnover. The sector with the highest grossing turnover (in millions of pounds) earned in 2001 is ranked number 1.

The stages of engineering and manufacturing a product

Most products that are engineered or manufactured follow a similar pattern from design, through production until they finally end up in the hands of the customer.

The stages of engineering and manufacturing are shown below.

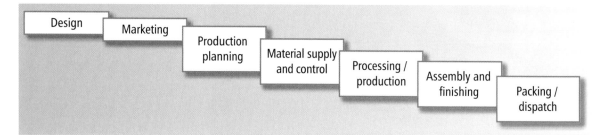

The first part of any production activity is the design stage. It doesn't matter if it's a pair of trainers, an oil rig or a DVD player, all need to be accurately designed and developed.

In the past, designing would be carried out on drawing boards by highly skilled designers and drawing office personnel. The use of computer programs is now very widespread, and the vast majority of final design work is carried out using specialist software.

This specialist software is often termed **computer-aided design** or CAD for short – we will discuss this later in the book.

Products are generally designed for two main reasons:

- to provide a functional solution to a problem (i.e. design a brand new item for a specific reason)
- to improve the performance or aesthetics of an existing product.

The process of doing research to find gaps in the consumer market is part of the **marketing** process. Information is often obtained through customer **questionnaires** and **feedback**. Have you ever bought a product and sent back a question card to the manufacturer with your thoughts on how good you think the product is? If the answer is yes then you have contributed to a marketing campaign.

Production planning and **material supply and control** are often closely linked. They are part of the process of organising a company to produce the right amount of product, in the right order, using the right materials and equipment. This is often done by specialist production planning personnel working closely with stock controllers and purchasing departments. Together they ensure that the right amount of product is manufactured at the right time.

Processing and **production**, **assembly** and **finishing** are also closely linked. These are the actual engineering and/or manufacturing involved in producing a product. This stage of production is often highly automated, accurate and economical. It is possibly the most interesting stage in the production process – quite simply, it's where things get made!

Processing and production involves the actual making of the various parts and components, while assembly is the process of fitting these individual parts together. Finishing is the process of cleaning up or applying final touches to a product – painting or varnishing, for example.

The final stages are **packing** and **despatch**. In these stages the products are securely wrapped up and then sent to the distributors and retailers who eventually sell the products to us.

Think IT THROUGH

1   Which sector would the following products fall into?
    - a microwave
    - a catalogue
    - a golf club
    - a chocolate bar

2   A worker is labelling boxes full of finished training shoes. What stage of the engineering/ manufacturing process do you think this is?

3   Which of the following sectors has the highest annual turnover in the UK?
    - textile production
    - metals and fabricated products
    - food products
    - transport equipment

The use of information and communications technology

Information and communications technology, or **ICT**, is as widespread in engineering and manufacturing as it is in the home.

Companies use ICT every day in their production processes, and without it customers like us would not have the range and quality of technically advanced products available today.

There is a good chance that you are familiar with the popular ICT packages such as wordprocessors, spreadsheets, databases and Internet sites, as these are often very well used in school, college or at home. We will come to these later, but first let's start with the ICT used to design a product.

CAD is used all over the world to design products, from housing estates to mobile phones. It can be used to make layout plans, component drawings and 3D models.

THE JARGON DRAGON

CAD – computer-aided design. It involves the use of computers to carry out design work that used to be carried out manually

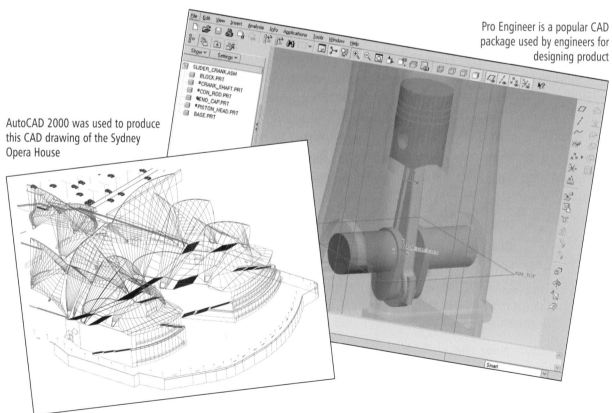

Pro Engineer is a popular CAD package used by engineers for designing product

AutoCAD 2000 was used to produce this CAD drawing of the Sydney Opera House

Even films such as *Shrek* and *Lord of the Rings* use an advanced type of CAD to construct their visual effects.

What is CAD?

You are probably already familiar with computer packages to help with design. Desktop publishing software, and graphics packages such as CorelDraw and Adobe Photoshop, are among the commercially available software used for computer design.

In industry, CAD generally refers to software packages that can produce high quality drawings and computer-generated models to exact specifications. Such packages enable the operator to draw objects very quickly, effectively and accurately. They can open, display and print drawings produced by other people that may have been sent on disk or by e-mail.

A popular package used for image manipulation is Adobe Photoshop

The main advantages of using CAD

Using CAD saves vast amounts of time and money. This means CAD has several key benefits:

- it produces drawings to a high standard and accuracy that is repeatable time and time again
- it can produce drawings more quickly than manual methods – this can lead to financial savings due to reduced labour costs
- it can carry out a range of drawing functions including 2D and 3D
- it is user-friendly and requires less manual skill to produce high quality drawings
- it is easy to open and modify existing drawings, which can then be printed, saved to hard drive, floppy disk or CD, or even e-mailed around the world

- standard parts such as screws, nuts and bolts can be pre-drawn and imported into drawings. This saves time, as the designer does not need to draw them from scratch each time he or she produces a new drawing.

Are there any disadvantages?

Yes, despite its many advantages CAD does have several disadvantages and limitations. They are:

- people may require special training to be able to use CAD – this can be expensive
- the software can be expensive to buy
- fewer people are needed to carry out drawing work because it is so much quicker than manual methods – this means fewer draughtsmen are required
- some CAD packages are not compatible with others, which may hinder the development of some products if different companies use different systems. A package called AutoCAD tends to be standard in many companies.

THE JARGON DRAGON

hardware – any part of the computer that you can physically touch, e.g. monitor, keyboard, hard drive, etc.

software – the programs that run on the computer. A well-known example would be Microsoft Word

What is required to run a CAD system?

Standard CAD packages can run on normal domestic computers, like the one you may have at home. More advanced packages might require more memory and higher processor speeds in order to run effectively.

The diagram below summarises the important pieces of **hardware** and **software** required to run a CAD system:

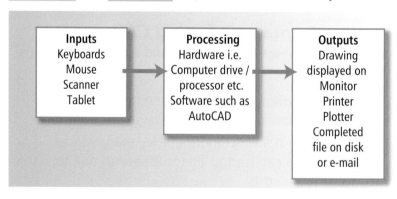

Inputs	Processing	Outputs
Keyboards	Hardware i.e.	Drawing
Mouse	Computer drive /	displayed on
Scanner	processor etc.	Monitor
Tablet	Software such as	Printer
	AutoCAD	Plotter
		Completed file on disk or e-mail

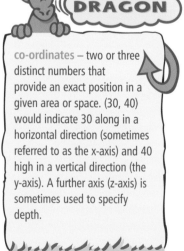

THE JARGON DRAGON

co-ordinates – two or three distinct numbers that provide an exact position in a given area or space. (30, 40) would indicate 30 along in a horizontal direction (sometimes referred to as the x-axis) and 40 high in a vertical direction (the y-axis). A further axis (z-axis) is sometimes used to specify depth.

How does CAD work?

Drawing using CAD is very similar to drawing by hand. The screen is your piece of paper and the functions (line, circle, polygon, etc.) are your drawing tools.

Most technical CAD packages use a system of **co-ordinates** to determine the exact position and size of features.

For an example of a CAD package, take a look at the screenshot of the house plan below:

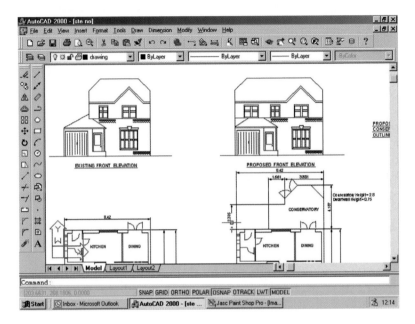

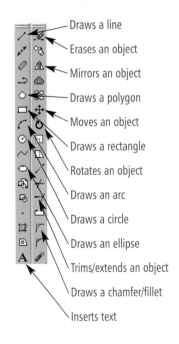

- Draws a line
- Erases an object
- Mirrors an object
- Draws a polygon
- Moves an object
- Draws a rectangle
- Rotates an object
- Draws an arc
- Draws a circle
- Draws an ellipse
- Trims/extends an object
- Draws a chamfer/fillet
- Inserts text

The **drawing area** is where the actual work is displayed, in two or three dimensions. This part of the screen represents the paper and it shows you what would be printed out. The drawing area can be changed to suit any paper size and the operator can zoom in very accurately without distorting the image. Likewise, any part of the drawing can be selected and printed – this is very useful when working with large products, for example vehicles such as cars.

The **drawing tools** are the basic commands used to create CAD drawings. The **tool bar** is shown to the right in more detail with descriptions of the most common commands.

The **command bar** is used to input commands using the keyboard. For example, to draw a line you would need to type in the command bar the command (in this case 'line' or 'L'), and the starting and ending co-ordinates.

e.g. Line from 30,40 (to) 100,50

This would result in the screen shown at the top of the next page – note the text within the highlighted command bar:

Many of the other functions or commands work in much the same way.

By using a variety of shapes and features with very accurate co-ordinates it is possible to design the vast amount of engineering and manufacturing products available today.

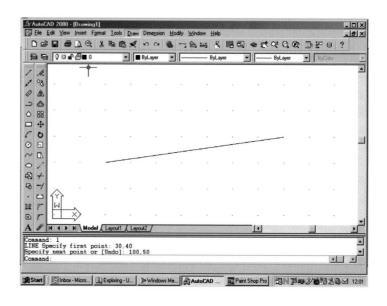

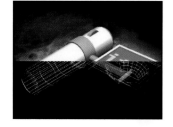

The finished product design can be displayed in two or three dimensions, as wire-frame or rendered images – modern packages can also animate objects to show multiple sides and faces in a range of angles and views.

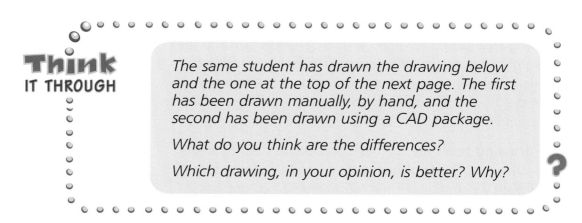

Think
IT THROUGH

The same student has drawn the drawing below and the one at the top of the next page. The first has been drawn manually, by hand, and the second has been drawn using a CAD package.

What do you think are the differences?

Which drawing, in your opinion, is better? Why?

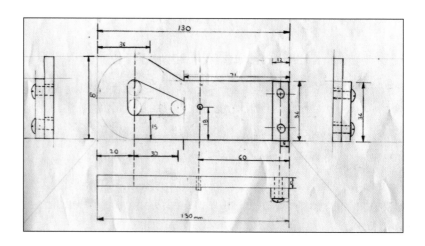

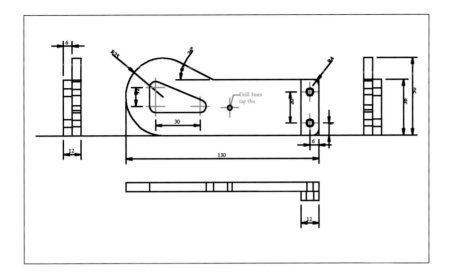

Introduction to CAM

The twentieth century saw massive technical improvements leading to the invention of many fundamental manufacturing techniques. Robotics, automation and computer technology were all developed in the last century and now contribute significantly to the range of products we have available today.

The list below shows how some of the key manufacturing and engineering techniques and processes were developed within the twentieth century.

- **1920:** first use of the word 'robot'
- **1920–1940:** mass production boom, first use of transfer machines
- **1940:** first electronic computer
- **1943:** first digital computer
- **1945:** first use of the word 'automation'
- **1952:** first prototype of a CNC machine
- **1954:** development of NC programming code
- **1957:** first commercially available NC machine tools
- **1960:** first use of industrial robots and CAD systems
- **1970:** first integrated manufacturing system
- **1970:** further development of CNC systems
- **1980s to present:** development of CAM techniques.

Adapted from: Serope Kalpakjian, 'Manufacturing Processes for Engineering Materials'

Some of the words above may not be clear to you – don't worry, they will be explained within the next few pages.

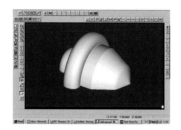

Computer-aided manufacturing

Computer-aided manufacturing (CAM) allows computers to be used to design, develop and manufacture products with very little effort compared to more conventional techniques.

This means a product can be designed on a computer using CAD and then transferred to a CAM package. It is also possible to draw a component from scratch with many CAM packages.

The computer-generated design can then be given further information that is required to process the product. The additional information could be:

- materials to be used
- equipment and machinery required
- sequence (or order) of operations to carry out manufacturing.

Once the software package has all the required information it can then produce a program that the manufacturing process can understand.

This program is termed **numerical control (NC)**, and as computers carry out the transfer link, the process is termed **computer numerical control (CNC)**.

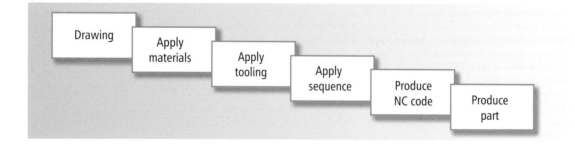

The manufacturing system is then used to produce the actual product or part of the product. In some cases it is used to make moulds or tools used in another process to manufacture the finished product.

Computer numerical control

Until the mid-1950s machinery used to cut and form was generally operated by two main methods.

- Manual operation (which is still widely used today) is carried out by very highly skilled workers who use various wheels, levers and controls to adjust the machinery to manufacture the required component. Measuring devices such as micrometers (see Unit 2) are used to control the accuracy and subsequent quality of the product. Altering and reworking are still very common to the process.

- For operations used to make a high volume of products, automatic operations tended to be more economical. These processes were faster and more reliable than manual methods but they did prove expensive to set up. This was because of the control systems, which used mechanical gears, cams and pulleys.

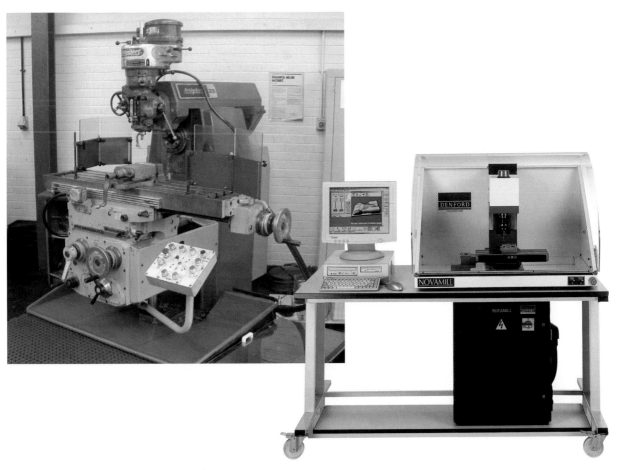

A traditional milling machine (left) and a CNC milling machine (right)

Numerical control changed all this. Numerical control used a program which described in exact detail the movement of the machine required to produce a part. The tooling would also be included, and the program could be used to produce identical parts over and over again.

In the early days the machinery didn't possess the memory capabilities that computers have today, and the program had to

A plasma-cutting process utilising a CNC system

be fed into the processor via punched paper – later plastic cassette tape was used.

Computer numerical control was a development that further improved the processing of products. By programming directly into the machine's memory using consoles, there was no longer any need to provide a tape copy of the program.

Operators could be given a drawing of the component and then, using a specific code, could program the machine to work automatically. The machine could then be set to work, allowing the operator to move (if required) to another workpiece or job.

This was a tremendous leap forward, and with the innovation of CAM, manufacturing high quality products to exact specifications was made even easier.

Manufacturing systems that are commonly linked to CNC include:

- drilling and milling machines
- cutting and welding operations
- bending and forming machines.

Typical products that use CAM techniques in their production include:

- scooters
- mobile phones
- games consoles
- trainers
- food packaging
- televisions

- CD players
- football boots
- water bottles
- office furniture
- mountain bikes
- shower gel containers.

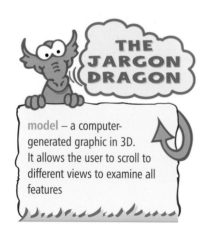

model – a computer-generated graphic in 3D. It allows the user to scroll to different views to examine all features

How does CAM work?

The exact operation of computer-aided manufacturing varies from system to system, but it generally follows the stages:

- generate a computer drawing or 3D **model**
- decide what processes are to be used
- tell the computer how to manufacture it, i.e. what tools and sequences are to be used
- if necessary, use the software to 'prove' the sequence before actually making the product
- generate the numerical code (NC) program
- send it to the machine
- set the machine going and make the product.

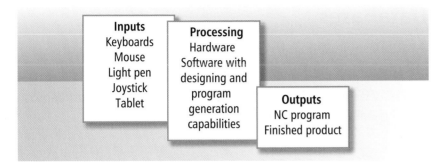

Inputs
Keyboards
Mouse
Light pen
Joystick
Tablet

Processing
Hardware
Software with
designing and
program
generation
capabilities

Outputs
NC program
Finished product

CAM has various advantages over more traditional manufacturing and engineering techniques. For example:

- it is easier to make technically complicated components
- it produces high quality products that are very accurate, time and time again
- it is more efficient than traditional methods, i.e. it is quicker and can produce high volumes of product
- programming can be carried out off-line in specific design offices and programs downloaded to machines in a different area
- operators do not need to be as highly skilled as those operating traditional processes.

The advantages of CAM to companies usually outweigh the disadvantages; however you need to note the following:

- it can initially be very expensive to set up CAM systems
- it can reduce the number of staff required in a company, leading to fewer employment opportunities
- as machine operation is largely controlled by technology, it can lead to a reduced level of skill among operators
- some systems can have compatibility problems with others, making file transfer between them difficult.

Think IT THROUGH

1 What are the advantages of CAD?

2 What are the inputs into CAD?

3 Are there any disadvantages of CAM?

4 What are the outputs from CAM?

5 Which of the following packages are not CAD packages? Paintshop Pro, Word, AutoCAD LT, Photoshop, Excel.

Databases

A database, in simple terms, is an organised collection of information (or **data**, as the information is often termed).

You may or may not be aware that your own details are kept on many databases right now:

- at school or college
- at the doctor's, dentist or hospital
- at the local video store
- at youth centres and sports clubs.

This data can often include personal information such as your name, address, telephone number, age or height.

In industry, companies use databases for several reasons:

- **information on employees:** including personal details, salaries and job descriptions
- **information on customers:** including past purchases, account numbers and personal details such as home addresses, e-mail addresses and phone numbers
- **information on suppliers:** including contact details, items supplied, delivery dates and cost
- **information on stock:** material quantities, tools and equipment available
- **information on finished products:** products complete and stored ready for delivery to customers.

Within a database the data is sorted into a number of **records**. Each record contains a number of **fields** and it is within these fields that the data is stored. For example:

Table1 : Table

Field Name	Data Type
Name	Text
Address	Text
Town / City	Text
Country	Text
Postcode	Text
Phone Number	Number
Job Title	Text
Company Department	Text

This is from a database program called Microsoft Access

The previous diagram shows the fields being set up within the **design view**. They can be modified to hold data such as text, number, currency, date, time, yes/no or even a hyperlink.

Once the fields are set up in the design view they can be viewed as a table (as shown below). It is at this point that data can be entered to form records – in this case the records are a company's employee details.

File	Edit	View	Insert	Format	Records	Tools	Window	Help

ID	Name	Address	Town / City	Country
1	Tickle John	11 Red Road	Lincoln	England
2	Knoxville Jonny	4 Blue Street	Edinburgh	Scotland
3	Rowland Kelly	9 Mauve Close	Cardiff	Wales

Another view which is often used is the **form view**. The same data is displayed, although in a more attractive and accessible way. One record at a time is displayed (in this case John Tickle), with the fields being spaced and formatted as required.

Buttons can be added to the form, which can be used to carry out common database functions such as adding and deleting records, finding a record or printing a record.

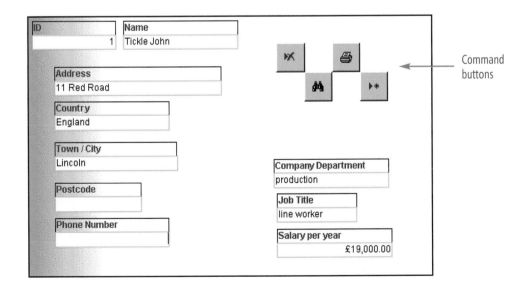

Database management systems
The **database management system (DBMS)** is a special program stored within the computer, which keeps track of

where the information is stored and indexed so that users can quickly locate the information they require.

Task performed by the DBMS include:

- adding new records
- deleting unwanted records
- amending records already contained within the database
- linking or cross-referencing records to find information
- searching or sorting the database to find exact data
- printing records.

A common database management system has already been mentioned – Microsoft Access.

Think
IT THROUGH

Using the Internet or an information technology textbook, try and find out about the Data Protection Act.

In particular, consider the following points:

- *In what year was it implemented?*

- *How does it apply to companies holding information on individuals?*

- *What do you think might happen if an Act like this was not in place?*

Example

Some companies might have several databases, each containing very different information used for specific purposes.

For example, a car manufacturer might have the following databases:

- a product database containing records of each vehicle model that has been manufactured. It might include engine size, colour and any extras fitted such as alarms, CD player or a sunroof
- a manufacturing database containing records of all the components and materials used during manufacture. This list is often referred to as a 'bill of materials' or 'parts list'
- a customer database containing records of all the dealerships that have purchased cars. This would include chassis serial numbers so that the manufacturer knows

exactly what has been supplied to the customer. If there was a problem with a specific car, this database could be used to locate and recall it

- a spare parts database containing records of all spare materials and components required for servicing and replacing defective parts. This database would include what part of the store they were held in, and the quantity available.

The task of the DBMS is to co-ordinate all this information so that data can be cross-referenced between databases.

Spreadsheets

A spreadsheet is simply an organised way of allowing a user to enter and display information, carry out calculations and in some cases perform other functions such as displaying charts and graphs.

Each spreadsheet file contains a number of **worksheets** – individual pages containing **cells**, which hold the data. The worksheet cells are formatted into **rows** (running across, shown in red below) and **columns** (running down, shown in blue).

The most popular spreadsheet used today is Microsoft Excel, shown below.

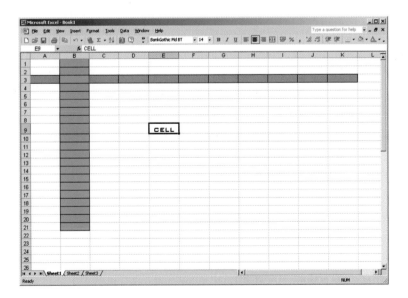

Spreadsheets are very useful for organising numerical data because the software is able to perform calculations – this explains their widespread use in the area of finance.

Popular applications of spreadsheets within manufacturing and engineering companies include profit/loss accounts, budget planning, project management and quality assurance recording.

One of the main advantages of spreadsheets is their ability to display data in a range of formats, including:

- table layout
- line graphs
- pie charts
- bar charts
- scattercharts (often used in quality assurance)
- area charts.

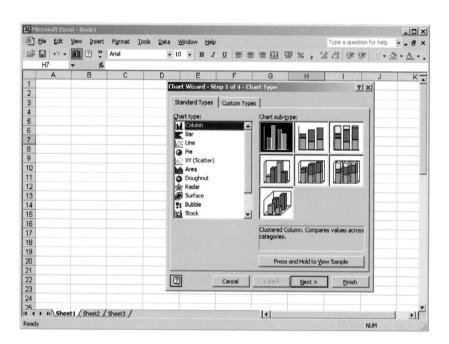

Other features of a spreadsheet include:

- you can insert, cut and paste information from other applications such as databases and wordprocessing packages
- formatting, such as changing row/column size, merging cells, adding colour and changing the font characteristics, is extremely easy
- unwanted information can be deleted and edited, copied or moved as required
- it is possible to search for key terms, filter and sort data into specific order – alphabetically, for example
- functions such as SUM, AVERAGE and IF can be used in calculations

- files can be transferred via e-mail to other individuals, departments or companies
- information from other worksheets can be used in calculations.

Example

Consider a fruit juice carton – before it is folded into shape it must first be cut to length.

A supplier of the material used to make the cartons has installed a quality system that ensures that the card is always cut to a specific length. The quality control department uses a spreadsheet to enter the cut-card length, analyse the results and to monitor if the process is 'in control'.

The **nominal value** is the ideal target value, which is manually entered by the QA (quality assurance) technician. The card is then automatically measured and the results inserted into the 'measured cut length' column of the spreadsheet.

The spreadsheet then calculates the average, which is then compared with the nominal value to produce a deviation value.

Any deviation below 0.5 mm is acceptable; hence this process, having a deviation of 0.02 mm, is in control.

The diagram on the next page shows the information displayed as a line graph to detail the fluctuations in the process – note the formulae for the simple calculations shown in the formulae bar.

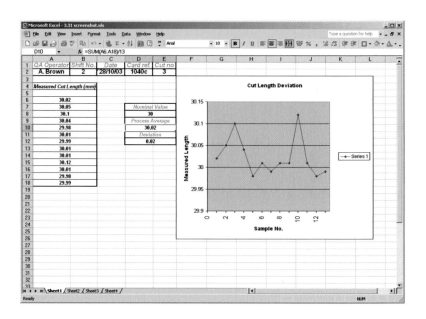

1 *Why would a company use a spreadsheet?*

2 *State an advantage of using spreadsheets.*

3 *Why would a company use a database?*

4 *With a database, what is the difference between cells, fields and records?*

THE JARGON DRAGON

infrastructure – the way in which communication is linked and organised

Telecommunications

Telecommunication is a term used to describe the **infrastructure** that allows a range of information to be transferred from a source to a destination. This information could be sound (for example, a telephone conversation), text (such as e-mail), graphics (web pages) or vision (TV programmes).

Telecommunications include communication by conventional wires (or lines), cables, radio, satellite and optical technology.

Devices such as mobile phones and modems for computer Internet access are common examples of telecommunications products.

Cables and wires

Traditional cables and wires are generally made up of copper insulation with a PVC coating. This is the type of wire used to carry electricity in your home – it connects all your plug sockets and light switches to the main electrical supply. It is also used to carry electricity to most home appliances – your microwave, toaster, TV and DVD player all use this type of wire.

Traditionally, all phone lines consisted of this type of cable, but many have since been replaced with more modern technology.

Coaxial cables have a clear plastic core surrounded by a PVC coating. They are used for a range of cables used in the broadcasting industry, including satellite and television leads. Telecommunications companies such as BT, Telewest and NTL also use this type of cable to connect your home phone line to the connection box in the street.

An **optical fibre** is a long, thin strand of very pure glass enclosed in an outer protective jacket – you may have seen these as Christmas decorations or in novelty gadget shops!

They work using light, which is internally reflected along the fibre at 200 million m/s (two-thirds the speed of light in a vacuum).

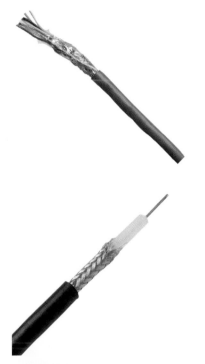

To establish a data link using optical fibre a transmitter/receiver unit must be set up at each end. This transmitter/receiver sends and recognises pulses of light from a light-emitting diode (LED) that is triggered by current from a computer interface.

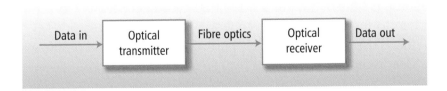

Fibre optics are also used for decorative products such as lights. The changing colours of the light travel down the glass fibres making them appear to change colour

Radio waves

You've probably heard of FM radio and AM radio – just look on your midi system – these are two examples of **radio waves**.

Radio waves are used to transmit music, conversations and other information though the air. They travel in waves that

hertz – the standard unit used to measure frequency, named after Heinrich Hertz, a pioneer in this field of science

oscillate many times per second. The number of oscillations (or cycles) per second completed by a radio wave is termed its **frequency**, and that frequency is measured in **hertz**.

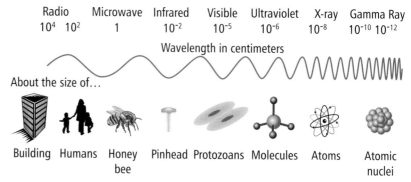

Although radio waves are light waves, we cannot see them; the human eye is unable to detect light at a frequency lower than that of visible light. Therefore even though radio waves are all around us they are invisible and cannot be seen.

Other light waves that cannot be seen include ultraviolet (used for sunbeds and sterilising in hospitals), infrared (used in remote controls and burglar alarms) and X-rays (used widely in medicine).

Example

If you are ever listening to Radio One and hear Chris Moyles, Jo Wylie or Sara Cox say 'you are listening to Radio 1 – 97–99 FM' the information they are giving you is the radio transmission frequency.

'97–99 FM' means that they are transmitting at between 97 MHz and 99 MHz. At 98 MHz, the transmitter at the radio station is transmitting at 98,000,000 cycles per second. To listen to these radio waves requires a frequency modulator (FM) – this is used to tune in to this particular frequency and produce a clear reception.

The main types of radio waves used for communicating are:

- **very low frequency (VLF):** used for long-distance communications over thousands of kilometres
- **low frequency (LF):** used in Europe for long-wave broadcasting
- **medium frequency (MF):** also known as medium wave, is used for broadcasting throughout the world

- **high frequency (HF):** HF or short wave has many applications, including broadcasting and long-distance (worldwide) communications
- **very high frequency (VHF)**
- **ultra high frequency (UHF):** used for both broadcasting and mobile communications, mainly over short distances (typically up to 100 km or more).

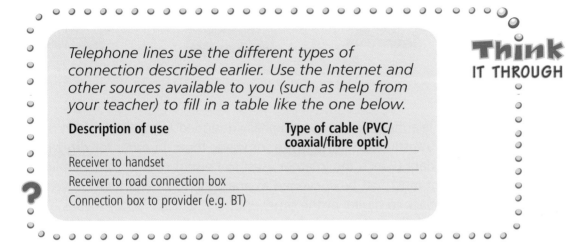

Telephone lines use the different types of connection described earlier. Use the Internet and other sources available to you (such as help from your teacher) to fill in a table like the one below.

Description of use	Type of cable (PVC/ coaxial/fibre optic)
Receiver to handset	
Receiver to road connection box	
Connection box to provider (e.g. BT)	

Think
IT THROUGH

The Internet

Take a look at the picture opposite. It's likely that although you would probably recognise film stars such as Brad Pitt or Russell Crowe, footballers such as Michael Owen, or pop stars such as Kylie, it's unlikely that you'll recognise this man. Yet he has had more impact on our daily lives over the past ten years than any of the other people mentioned above. We'll come to how he has achieved this later on.

Although the terms 'Web' and 'Internet' are often used together, they actually have two very different meanings.

The **Internet** is the global network that links computers worldwide. It carries data and makes the exchange of information possible.

The **(World Wide) Web** is a subset of the Internet. It is a collection of inter-linked documents that work together using common codes and language. This enables it to display material, which you know as web pages.

The Internet, therefore, consists principally of the hardware such as the modems, cables and wires that link our computers together as shown at the top of the next page.

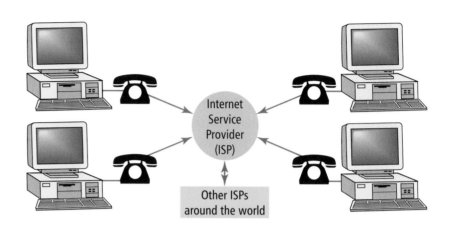

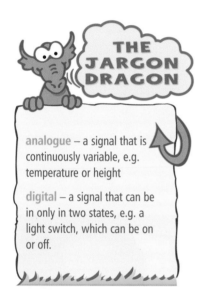

THE JARGON DRAGON

analogue – a signal that is continuously variable, e.g. temperature or height

digital – a signal that can be in only in two states, e.g. a light switch, which can be on or off.

Telephone lines were originally designed for speech, which is transmitted in **analogue**, or wave, form. In order for **digital** data (which the Internet uses) to be sent over a telephone line, it must first be converted to analogue form and then converted back to digital at the other end. This is achieved by means of a modem (MOdulator/DEModulator) at either end of the line.

The Internet has actually been around since 1969 where it was called the ARPANET – this was an acronym for the Advanced Research Projects Agency Network – and consisted of four computers. It was primarily used by scientists to transfer technical information from one to another.

It wasn't until 1990 that the Internet and WWW as we know it really took off, all thanks to one man. In March 1989, Tim Berners-Lee, an Oxford graduate and European physicist, proposed the idea of the World Wide Web as a means to better communicate research ideas amongst members of a widespread organisation. Berners-Lee (whose picture is shown on the previous page) went on to create the codes that we use today, and what's more he did it all for free. He invented the Web and simply gave it to the world.

The code used to link all the documents is called **hypertext transfer protocol (HTTP)**. This is a way of organising information so that both the sender and the receiver can understand it.

Websites are made up of a collection of web pages. These individual web pages are written in 'hypertext markup language' (HTML). The HTML tells Web browsers such as Navigator or Explorer how to display the various elements of a web page.

For example, this is the Sony web page as it appears to you:

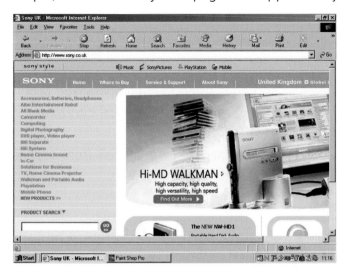

This is the HTML used to make up the page:

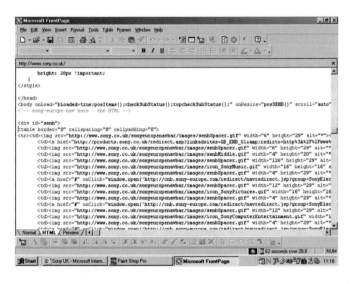

Internet service providers (ISPs) are responsible for enabling us to use the Internet – they are companies such as BT, AOL, NTL and Virgin.

The ISPs assign each computer a specific number called an **Internet Protocol (IP) address**. This is used to distinguish between individual computers using the net. Depending on the ISP you use and the contract that you have, you may have a permanent IP address or a different (dynamic) address each time you use the net.

Every web page on the Internet has its own unique address, known as a **uniform resource locator (URL)**. The URL tells a browser exactly where to go to find the page or object that it has to display. A well-known URL is http://www.ask.com.

How the Internet helps business

So how does the Internet help companies and businesses today?

Companies today generally use the Internet and WWW for various reasons. They include:

- marketing and advertising their company and products
- providing contact details and technical information to customers
- searching for and buying raw materials from suppliers
- selling products directly to customers.

Example

Take a look at the website below. It was designed by a GCSE engineering student and represents a fictional engineering fabrication company.

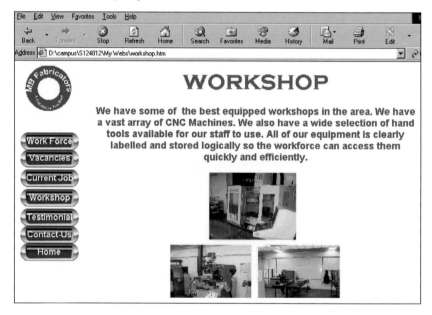

It features a homepage with company information, a logo and photographs. Within the homepage there are hyperlinks to other web pages:

- a staff organisation chart providing information on the staff and their roles and responsibilities
- a product page advertising the range of components fabricated in the past – in this case a screwdriver, hacksaw and mallet
- a manufacturing page showing the workshop and links to other sites
- a page about quality and how the company is responsible for producing quality products made to exact specifications.

Think IT THROUGH

? Take a look at any company website. Note down the different pages, how they are linked together and what information they contain. Consider why each page contains that information and what the company is trying to achieve by placing it there.

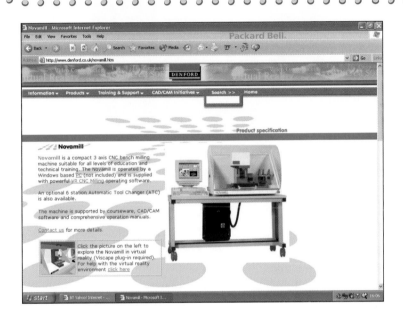

Electronic mail

Electronic mail, or **e-mail**, is arguably one of the most important developments in communication since the evolution of the mobile phone. It allows you to send messages, letters, documentation and various files by typing in the receiver's e-mail address and simply hitting the send button.

E-mail can be sent worldwide using the Internet or within an organisation by using an **intranet**.

Some advantages of e-mail include:

- a message can be sent to each person in a group of individuals at the same time
- at its quickest, e-mail takes only a few seconds to arrive at the recipient's inbox
- other files can be attached to a message (such as Word documents, spreadsheets and images)
- messages are saved automatically, and can be copied and printed for reference.

THE JARGON DRAGON

intranet – an internal network of computers, usually within a business, school, college or other medium to large organisation

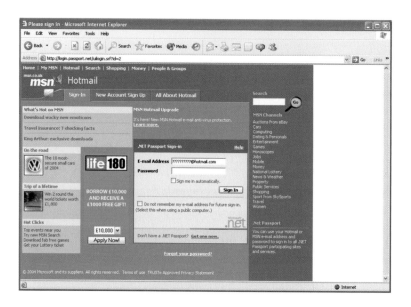

A popular way to send e-mail is by using the World Wide Web, for example Hotmail. You may use Hotmail to e-mail your friends, but it has many of the functions of more commercial e-mail packages. Hotmail allows you to send and receive messages, save and delete messages and attach documents such as JPEG pictures.

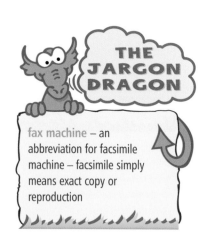

THE JARGON DRAGON

fax machine – an abbreviation for facsimile machine – facsimile simply means exact copy or reproduction

Before e-mail, engineering and manufacturing companies had to correspond by typing memos, sending information by external/internal post or by using **fax machines**. These techniques have several disadvantages:

Depending on the quality of delivery, posting information can take from an hour up to several weeks. Think of a company department based in Japan having to send a technical drawing to another department based in Scotland – this may take a few weeks to arrive! If the drawing was required urgently then this could cost the company valuable time and money.

While fax machines are much quicker than using the post, they depend largely on the quality of:

a the original document being sent

b the equipment being used.

Both these need to be of good quality in order to ensure good quality reproduction of the original document.

Companies can now use e-mail to arrange meetings, send minutes of meetings, issue memos, and send technical information, engineering drawings and quality information.

New materials and components

Introduction

The ongoing development of modern materials and components is closely linked to the engineering and manufacturing of exciting new products.

The development of materials dates way back, before 4000 BC, when metallic elements such as gold, tin and copper were first used for jewellery. The arrival of civilisations such as the Egyptians, Ancient Greeks and the Roman Empire further advanced the development of materials.

The Egyptians had a well-known association with ceramics – just think of the vast stone building blocks making up the pyramids. The ancient Greeks also used many ceramics, for fine pottery and porcelain.

The Romans used a diverse range of materials – metals such as lead, wrought iron and bronze were used in the military for weapons; gold and silver were used for jewellery; and ceramics such as marble and stone were used to create the vast structures, amphitheatres and buildings making up Ancient Rome. There is even evidence of an early composite being used, i.e. concrete (a combination of stone and cement), to provide strength within buildings.

In the Middle Ages the development of many metals continued. These included iron, bronze, steel and zinc for metalworking, armour and coins.

The Industrial Revolution (1750–1850) was a key part of history for engineering. Many processes were developed around this time, such as extrusion, rolling, drawing and casting, all to process new steels and irons available due to advances within materials science. The first **polymers** (plastics and rubber) were also developed around this time.

Arguably the most widespread development of materials occurred within the latter half of the twentieth century (1940 onwards). New metals were developed, such as aluminium alloys (used to make cars, aircraft, soft drinks cans, etc.); advanced polymers such as PVC and ABS; composites such as glass fibre; and new micro-electronics leading to advanced integrated circuits, an example being the Pentium microchips used so commonly today in computer design. Each of these new materials is discussed within this section.

Metals

Metals are usually combinations of metallic elements found naturally here on Earth. Common metallic **elements** used to make natural metals include gold, silver, tin, lead, iron, zinc and copper.

Alloys are metals that combine several different elements to make one material – good examples are brass (which is a combination of copper and zinc) and bronze (a mixture of copper and tin).

As discussed in Unit 2, metals are extremely good conductors of electricity and heat, are strong and yet can be shaped, which accounts for their extensive use in engineering and manufacturing.

Many metals, such as gold, tin and copper, have been used for thousands of years. Others, however, were developed quite recently. These include the following.

Aluminium alloys

Aluminium alloys possess high strength-to-weight ratios and have good resistance to **corrosion**. Aluminium alloys are very common and can be found as food packaging (cans and foil), in transport such as cars, aircraft and motorbikes, in buildings and as consumer goods such as furniture and kitchen appliances.

Magnesium alloys

Magnesium alloys are used, for example, in aircraft, golf clubs, printing machinery components, power tools and bicycles. Magnesium is often used as an alloy for parts that require good strength and are lightweight.

Nickel-based alloys

Discovered in 1751, nickel is a major alloying metal used to increase strength, toughness and resistance to corrosion.

Nickel alloys are used in a diverse range of products, including food-handling equipment, aircraft engines, nuclear power stations and coins.

Titanium alloys

Titanium alloys tend to be expensive and are therefore generally found only in very high performance products such as Formula One racing cars, aircraft engines, petrochemical and submarine parts.

Like many of the other alloys, titanium has high strength-to-weight ratio and excellent corrosion resistance.

The extreme strength and flexibility of alloys like titanium make them ideal for use in hostile environments such as outer space

ductility – the ability of a material to be pulled and formed into a desired shape without breaking or fracturing

Stainless steel

Developed in the twentieth century, stainless steel is a combination of steel (iron and carbon) with other alloying elements such as nickel, molybdenum, manganese, aluminium and, most importantly, chromium.

Often used to make cutlery, kitchen sinks and domestic utensils (sieves, whisks, drainers, etc.), stainless steel has an excellent resistance to corrosion, good **ductility** and high strength.

Shape memory alloys

Shape memory alloys are metals that can be formed into a shape at room temperature and will retain that particular shape. Once heated to a higher temperature, the material will 'remember' its original shape and deform back.

A combination of titanium and nickel produces a shape memory alloy.

fatigue – a form of material failure that is often caused by repetitive, cyclic stress – it often causes cracking of a component

Today and the future

Many of the developments in materials have been fuelled by the aerospace industry where it has been necessary to develop materials with very specific properties. These properties include high strength-to-weight ratios, making them very strong yet light, and a good level of resistance to corrosion and repetitive stresses such as **fatigue**.

Technology today allows us to produce a specific metal by combining various elements, with the finished metal having the desired characteristics of each individual element. This 'tailoring' allows batches of a unique metal to be made, specifically designed for maximum performance within a product.

Polymers

Polymers include the widely used materials plastic and rubber.

Many polymers are organic compounds based on carbon, hydrogen and other non-metallic elements. They are characteristically easy to shape, possess low mass, and many are extremely flexible.

Some of the numerous products manufactured from polymers include food and drink containers, packaging, housewares, clothing, paint, toys, computer products and car parts.

The first plastics were made from animal and vegetable products, **celluloid** (used in photographic film) being an early example. The first artificial, or synthetic, polymer was developed in 1906. It was called Bakelite and was used extensively to make telephone casings.

In 1920, the first modern plastic material was developed. Scientists extracted the raw material required to make plastics from **fossil fuels** such as coal and petroleum. Ethylene (used to make a popular plastic, polyethylene) was the first of such materials to be made, by reacting first coke and methane and then combining the product with hydrogen.

Put simply, polymers are long chains of molecules made by linking and cross-linking small repetitive groups of atoms, known as **monomers**.

The term 'poly' derives from the Greek meaning 'many' – hence many units of atoms.

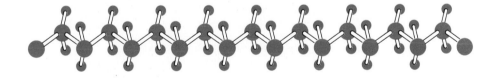

Plastics are often sub-divided into two groups – **thermosetting** and **thermoplastic**. Thermosets retain their shape upon reheating while thermoplastics become pliable and can be reset. Thermoplastics are the most commonly used plastics for domestic products.

Popular examples of thermoplastics are:

- **PVC (polyvinyl chloride):** used to make window frames, insulation for wiring, floor tiles and garden hoses

- **nylon:** used to make bearings and gears which are often found in electrical equipment with moving parts – for example, printers and video recorders
- **polycarbonate:** used to make safety helmets, lenses and as a base for photographic film
- **acrylonitrile-butadiene-styrene (ABS):** used for car parts, toys and as casings for electronic products such as digital cameras, CD players and mobile phones
- **polyethylene:** used to make drinks bottles, toys and clear film used as food wrapping
- **polystyrene:** used as packing (in its foam state), toys and CD cases.

Which plastics are used in the construction of a motorcycle crash helmet?

Examples of thermosets are:

- **silicone:** used to make fillers and sealants for bathroom furniture
- **polyesters:** used to make chairs, car parts and as a base for glass fibre
- **epoxies:** used to make sinks and shower trays and electrical mouldings such as plugs and sockets.

Composites

Composites are combinations of different types of materials mixed together. The final material has the best characteristics of each of the original materials.

A common group of composites is also called **'reinforced plastics'**. These materials are made by placing dispersed fibres within a selected plastic material known as the **matrix**. Fibres commonly used include glass, carbon and boron.

Glass fibre is a familiar example, in which fibres of glass are embedded within a polymer. This type of reinforced plastic provides the material with the strength of the glass and the flexibility of the polymer.

Carbon fibre is also often used to make products that require flexibility yet high strength, for example fishing rods and golf clubs.

Concrete is another example of a composite material. Cement; sand and gravel are combined to make the final product.

Think IT THROUGH

Identify the type of material and name of material for the products within the table below. For example, the first product is a can of Fanta. This is made from a metal called aluminium.

Product	Type of material	Name of material
Fanta can	Metal	Aluminium
Car tyre		Vulcanised rubber
DVD player casing		ABS
Digital camera casing		
Adjustable spanner	Metal	
Safety hat		
CD cover	Polymer	
Mountain bike frame		
Car engine casing		
Snowboard		Glass fibre

Selection of materials

In conclusion, with so many new materials available, how do engineers actually select a material for a particular function?

Generally, they apply the following rules:

1 The properties of the materials, such as hardness, strength, weight, ductility and conductivity, are usually the first consideration. For example, a computer casing needs to be rigid, with good resistance to electricity and heat; hence a plastic such as ABS is often used.

2 Cost and availability are also important considerations. A readily available material such as iron makes carbon steel relatively cheap. However, a metal such as titanium, which is less common, has a higher price tag. The mass processing of plastics such as nylon and PVC make these materials also relatively inexpensive.

3 When consumers buy products (especially those for domestic use) they are often influenced by their appearance – also called **aesthetics**. Materials such as plastics are very easy to colour and naturally have a smooth finish. They are therefore commonly used on many domestic appliances, children's toys and electrical equipment.

In addition to these general rules, manufacturers take into account many other factors when choosing materials. These include:

- the cost of the manufacturing processes to be used, for example casting, machining and joining
- whether the material can be processed by a particular method
- the disposal and recycling of waste materials, especially in the chemical and biological, electronic and automotive industries.

Electronic components

Introduction

Advances in electronic components have made possible the development of many consumer products we take for granted today. As electronic components become increasingly more complex and miniature in size, newer, improved products are constantly hitting the marketplace.

Think of the improvements in mobile phones and games consoles over the past few years. These have been made

possible by the tiny components transforming stored energy
from a battery or power cable into the sound waves used in
your phone or the video game you see on your TV screen.

Some of the most important of these components are described
below.

Resistors

A resistor is a device that doesn't allow current to flow as well
as a normal length of wire. It resists the flow of current, hence
the name. Resistance is measured in **ohms**. The higher the ohm
value, the more it resists the flow of current. Resistors come in
values from 0.1 ohm to 10 mega ohms (ten million ohms).

Capacitors

A capacitor is an electronic device used for storing electric
charge. It works in the same way as a tank used to store water
– when needed, it releases the required amount of charge into
the circuit.

The units of capacitance are **farads**. A farad is a very large
amount of capacitance, so submultiples such as microfarads
(10^{-6}) and picofarads (10^{-9}) are used.

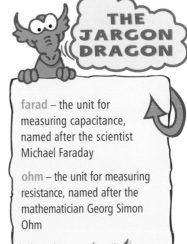

THE
JARGON
DRAGON

farad – the unit for
measuring capacitance,
named after the scientist
Michael Faraday

ohm – the unit for measuring
resistance, named after the
mathematician Georg Simon
Ohm

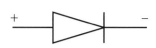

Diodes

A diode is a semi-conducting device, which allows electricity to flow in only one direction but not the other. It is like a one-way valve for electricity.

By looking at the symbol, you can see that it has the shape of an arrow. The electrical current will flow in the same direction as the arrow.

There are special terms used for each side of the diode – the positive side of the diode is called the **anode**. The negative side is called the **cathode**. Current always flows from the anode to the cathode.

LEDs

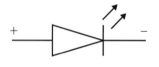

LED stands for light-emitting diode. An LED is a diode with a unique property – when electricity flows through it, it emits light. The LED requires only a small amount of voltage to operate and so it is used in electronics more often than light bulbs.

The circuit symbol for an LED has an anode and a cathode, as on a regular diode, but you know that it is an LED from the little arrows pointing away from it. This represents light energy being emitted.

Batteries

A battery is made up of two or more cells. Batteries supply voltage to a circuit which is used to drive the electric current around. They have two terminals; one terminal is marked (+), or positive, while the other is marked (−), or negative.

Electrons collect on the negative terminal of the battery. If a wire is connected between the negative and positive terminals, the electrons will flow from the negative to the positive terminal.

Inside the battery itself, a chemical reaction produces the electrons. The speed of electron production by this chemical reaction (the battery's **internal resistance**) controls how many electrons can flow between the terminals. Electrons flow from the battery into a wire, and must travel from the negative to the positive terminal for the chemical reaction to take place. That is why a battery can sit on a shelf for a year and still have plenty of power – unless electrons are flowing from the negative to the positive terminal, the chemical reaction does not take place. Once you connect a wire, the reaction starts.

Variable resistors

Variable resistors have a dial or a knob that allows you to change the resistance. Volume controls are variable resistors. When you change the volume you are changing the resistance, which changes the current. Making the resistance higher will let less current flow so the volume goes down. Making the resistance lower will let more current flow so the volume goes up.

Wire

Wire is used to electrically connect circuit parts, devices, equipment, etc. There are various kinds of wiring materials. Materials with good electrical conductance are always used, for example copper. On printed circuit boards (PCBs) connections are made by copper tracks which connect all the components. On small circuits these tracks are incredibly small and intricate.

Lamps

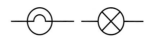

Lamps are used as indicators and for illumination. The most common type used has a filament coil which glows white hot when a current passes through it. The filament is in a space filled with an **inert gas** to aid the light output and to prevent corrosion of the filament. Bulbs use more power than LEDs but give a higher light output.

They are fixed into place with either a bayonet or screw cap fitting – both are very common.

Motors

A motor is a device for changing electrical energy into movement. The turning power (or **torque**) of a motor can vary greatly. Generally, small motors have low torque whereas large motors usually have a high torque rating. Motors come in AC and DC versions. DC types of 3–12 volts are the most suitable for general use.

Transformers

A transformer is made from two coils, one on each side of a soft iron core. It can decrease the voltage (called a step-down transformer) or increase the voltage (called a step-up transformer).

Transformers are widely used as part of the power supply to electrical equipment. Look on your printer or games console power lead – the larger plastic part houses the transformer.

Transistors

In electronics, signals sometimes need to be **amplified** (increased in magnitude). Take, for example, a mobile phone – the signal from the aerial is relatively weak and needs to be boosted so that it can be heard on the speaker. The component used to do this is a transistor.

There are two basic types of transistor: the n-p-n type and the p-n-p type. These vary according to the configuration of the materials used to produce them, which in turn affects the direction of current flow.

A transistor has three main parts: the collector (of the current), the base (which amplifies the signal) and the emitter (of the amplified signal). These are the three 'legs' of the transistor, as shown in the photograph below.

Integrated circuits

Integrated circuits, or ICs, are extremely small circuits containing resistors, diodes, transistors, connections and other components. These circuits are contained within a single 'chip' of a semi-conductor material, typically silicon. ICs can be made using chemical and microphotography techniques that are carried out automatically at a relatively small cost.

Before the invention of ICs in around 1960, circuits were much larger and more expensive, and consisted of various standard-sized components. ICs revolutionised the electronics industry and can be found in just about any electronic product – phones, calculators, digital cameras, computers, DVD players and MP3 players all contain integrated circuits.

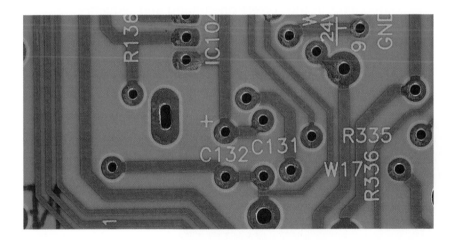

PCBs

Printed circuits boards (PCBs) are used to connect and hold electronic components together in one rigid structure.

Just about every electrical product manufactured today has a PCB inside it – from the relatively simple circuits in kettles or toasters to the highly technical models used in computer equipment.

PCBs are usually manufactured from board or plastic composites. The most common material used is glass fibre because of its excellent strength, resistance to electrical current and corrosion.

PCBs are made using a combination of photographic and chemical processes. These are used to print the copper tracks, which connect the components. A solder-resistant layer and text notation is usually also added via a printing process. A punching or CNC process is then used to achieve their final shape.

PCBs can be manufactured relatively easily in large quantities, which make them an invaluable part of many products in the marketplace.

Automated assembly of printed circuit boards

Components are usually applied to PCBs using highly sophisticated pick-and-place robotics, auto-insertion machines or by manual labour.

PCBs that are assembled using pick-and-place robotics are often described as using 'surface mount technology'. The robots are programmed by an operator 'teaching' a microcomputer (within the machine) where and how the parts should be placed. The machine then loads the PCB, repeats the program and applies the components in the correct orientation and sequence.

Auto-insertion works in much the same way although it is used for components with legs. The legs are pushed into holes (previously drilled within the PCB) and are bent round and cut to size to secure a good fit to the PCB.

Systems and control technology

Introduction

Until the 1950s most manufacturing operations were carried out using traditional machinery such as milling machines and presses, which were worked manually. As new complex products were developed it meant the operators struggled to make parts that were exactly alike, and often several trial-and-error attempts had to be made to get it right. The word 'automation' (meaning self-acting) was derived in the mid-1940s by the US automobile industry to indicate 'automatic handling of parts between production machines'.

These highly mechanised machines often required people with special expertise to set them up, and they were used only for producing very high volume, mass-produced products such as cars. This continued for several years until the development of the first computers.

During the 1960s the introduction of computers was a major factor in the development of automation. Computerised technology led to computer numerical control (CNC), automated material handling, industrial robots, computer-aided design (CAD), computer-aided manufacturing (CAM), computer-aided engineering (CAE), and computer-integrated manufacturing (CIM).

CAD, CAM and CNC have all been discussed earlier in the book. Now we shall look at the other terms listed above.

Computer-aided engineering (CAE) is a term commonly used to encompass CAD, CAM and CNC. It therefore refers to the design, 3D modelling, scientific analysis and processing of a product.

Materials handling

Materials handling/transfer is the process used to transport parts and materials between workstations and assembly points.

Conveyors

A common type of transfer machine is the **conveyor** – you may have seen conveyors in the airport, at the check-in desk or when collecting your luggage. You may also have seen them at the check-out at the supermarket.

Engineering and manufacturing companies use conveyors to move parts around, therefore reducing the need for manual handling. Common types of conveyors are belt, roller, slat and overhead. They are usually powered using an electric motor connected to a drive mechanism.

Belt conveyors are generally used on assembly or packing lines. They are also often found in the food and pharmaceutical industries, which require very clean manufacturing environments.

Slat conveyors are often used where rapid speed and accuracy are required. Again, the pharmaceutical and food industries are the largest group of customers for this type of conveyor. The biggest advantage is that they can open up, which allows them to move around bends, and they can be washed down for hygiene purposes.

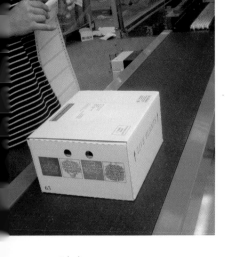

A belt conveyor

A slat conveyor

Roller conveyors are sometimes called gravity conveyors. They are generally used when transporting packaging such as plastic crates or cardboard boxes; consequently they are often found in the despatch department of a company.

A roller conveyor

Unlike other conveyor types, they usually do not rely on electrical power. Instead, the products are 'freewheeled' along the conveyor or slid down a slight drop, therefore using gravity to move the item.

Overhead conveyors are primarily used to save space. Most are electrically powered, but some use low-friction, free-running designs.

Overhead conveyors are widely used in the automotive industry, by cycle manufacturers, mail order companies and clothing manufacturers.

An overhead conveyor

Remotely operated/automated guided vehicles

Remotely operated vehicles (ROVs), sometimes called automated guided vehicles (AGVs), are normally used in larger manufacturing and engineering environments. They can be found predominantly in the aerospace, automotive and paper and board industries for handling materials and parts, and carrying out warehouse operations.

They are basically robot vehicles that move around automatically, without the need to be 'driven' by human operators. ROVs normally navigate by following a wire embedded in the floor of the factory, or by using a laser which detects their position based on specific points within the factory. They often form part of a complex manufacturing system in which they ensure that parts and materials arrive just as they are needed within the manufacturing process.

An example of an ROV is shown below:

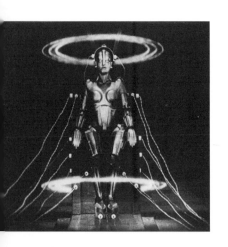

Industrial robots

If you were asked, 'What is a robot?', the chances are that you would be thinking about one like that shown on the left.

Although this is a fictional robot, robots in industry are very real.

The word 'robot' is simply a Czech word meaning 'worker'. First used in the early 1960s, early industrial robots were used to carry out hazardous operations such as handling toxic and radioactive materials. Industrial robots have since developed into multifunctional workers able to work with very high accuracy and repeatability. They have been defined as a 'reprogrammable, multifunctional manipulators, designed to move parts, tools or other specialist devices'.

Robots developed to use computer technology, and in 1974 the first robot controlled by a microcomputer was introduced. Robots are now very widespread in industry, and are used in:

- materials handling
- spot welding
- machining
- spray painting
- automated assembly.

Robots' components

The following diagram shows an articulated robot – this type of robot is very common in industry.

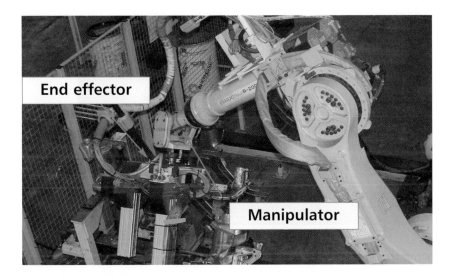

The **manipulator** is also called the 'arm and wrist'. This is because it provides motion similar to that of your arm and wrist. In the past, movement was made possible by mechanical devices such as gears and links; more modern robots use electric motors for more accurate control.

The **end effector** is basically the end-of-arm tooling. Depending on the function of the robot, it could be equipped with grippers, hooks, spray guns, power tools or more specific equipment such as materials handling fixtures.

The **power supply** provides the movement of the manipulator and end effector. The power supply for the robot in the photograph is electricity, although in some cases pneumatic (air pressure) and hydraulic (oil) power are used.

The **control system** communicates the instructions used to move the robot to its various components. This is essentially the brain of the robot, which tells the other components where and when to move.

The most common type of robot used in industry uses a control system termed 'playback'. This type of control allows the operator to 'teach' the robot the sequence of movements required to complete a processing operation. The robot then remembers these movements using co-ordinate points in various axes. When all the stages are complete, the robot can 'play back' the program and hence repeat the production task as many times as needed.

Programmable logic controllers – PLCs

PLCs are used in industry to control just about everything, from materials-handling conveyors to highly automated assembly equipment.

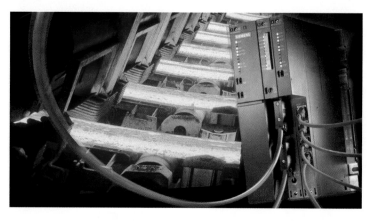

Example

A switch is required to turn on a conveyor belt motor and sound a bell to safely alert production staff after 10 seconds.

A simple external timer and bell could achieve this. However, if this switch had to control 20 conveyor belts then 20 timers would be needed – this would obviously require more setting up and additional cost to the company.

If a PLC was used, then it would use a program to switch all the motors on at the correct time. This program would control an internal set of instructions held within the PLC (in the CPU) called logic.

PLCs allow the programmer to wire up as many inputs (switches, sensors, etc.) and outputs (motors, lights, etc.) as required, and to control them by programming the unit with the required instructions or logic.

Before the introduction of PLCs in the late 1960s, such machinery was controlled using large, manually wired cabinets containing banks of wires, switches, relays and other components.

PLCs provide the same function but in a much smaller package. Roughly the size of a loaf of bread, the PLC contains the following key parts:

- **Central processing unit (CPU):** this controls all the PLC's activities in response to a program held in the memory.
- **Input cards:** these are the connections to external devices such as switches and sensors. They provide an input signal to the CPU.

- **Output cards:** these are also connections to external devices, but this time to the output components such as motors, lights or a siren.

Some advantages of PLCs are that:

- reprogramming is extremely easy compared with the manual rewiring of an entire cabinet
- they have extremely good repeatable operation – they don't seize or corrode like conventional components
- faultfinding can be carried out very easily.

Check out the website www.PLCS.net for more information.

Computer-integrated manufacturing (CIM)

CIM is the term used to describe the computerised connection of all aspects of design, planning, processing, distribution and management. It can be summarised in the diagram below. CIM is an integral part of many engineering activities and is vital in producing many of the quality products we use today.

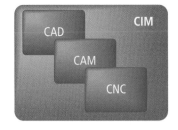

CAE (computer-aided engineering) can be described as an integrated process. The drawing is produced using CAD and then processed using a CAM package. The completed CAM file is then downloaded to a CNC machine which physically produces the product.

The table below provides an example of the use of CIM for each of the engineering sectors:

Sector	Example of CIM
Engineering fabrication sheet	The use of CAD, CAM (CAE) and CNC to cut metal using laser and plasma-cutting equipment
Mechanical and automation	The control of robotics and ROVs
Electrical, electronic, computer and telecommunications	The design and control of the computer-controlled pick-and-place robotics used to place tiny electronic components on PCBs
Textiles and clothing	The CAE used to develop clothing pattern templates and then cut them out using specialist equipment
Food and drink	The control of highly complex conveyor systems used to transport food goods and ingredients
Printing and publishing	The use of desktop publishing linked to specialist photo equipment used to produce the printing press plates (see case study)

1 What is an automated production line?

2 State three advantages of using robotics.

3 On a robot, what is the manipulator?

4 What is a PLC?

5 What is CAE and what ICT does it involve?

6 Describe how CIM is used in engineering and manufacturing.

Nissan

Imagine a place – the size of a small town – where workers drive around on electric vehicles that hardly make a sound.

In this place vehicles can guide themselves and shuttle from place to place, picking up parts that automated machines and humans need to carry out their jobs.

Imagine a place where humans work alongside hundreds of robots. They are all working together to meet a specific objective.

Some of the robots even think for themselves.

This is the Nissan factory in the north-east of England.

Covering 3 million square metres, Nissan UK is based near Washington, employing around 5000 people and pumping £400 million a year into the local economy.

The plant is responsible for producing three different models of car for all of Europe and is capable of producing a car every 60 seconds.

With every individual car having approximately 5000 components, human workers need a little help.

In this case the help is in the form of nearly 1000 industrial robots carrying out a range of activities, including moving parts such as doors and bonnets, welding using high-precision tooling at extremely high temperatures, and spray painting.

case study

Nissan

Working at high speeds of up to 2 m/s (metres per second), they meticulously carry out their functions with an urgency that makes them seem almost alive. Often they work in very close proximity, passing parts to each other, with each robot completing its task and simply passing on the finished job to the next workstation.

At Nissan the robots are used to carry out work that requires heavy handling or operations – work that causes humans fatigue or repetitive strains.

As Chris Purnell (Training Controller) at Nissan states: 'We are getting much better at identifying operator concerns and using this type of technology to assist.'

The introduction of this technology allows human workers to be repositioned on operations that require adaptability and intricacy – operations in which the introduction of automation such as robotics would be uneconomical. Chris continues, 'We are not trying to replace people, robotics allows us to relocate workers to environments where they can be utilised fully'.

These robots aren't what most people would expect. Take for example the Fanuc 2000I – it stands over 3 m high when at full stretch and could punch straight through a car roof with ease. It can be programmed for a range of operations and is used primarily at Nissan to carry out spot welding on the car body shells. The electro-operated servo-welding guns used as the end effectors can be adjusted to the perfect welding distance. This ensures a high quality weld, and also preserves the life of the copper welding electrodes used to supply the high current to the weld.

Electrically powered, the control units for these robots are roughly the size of a fridge and use electric servo-motors to transfer movement to the manipulator and wrist.

Programming is carried out using handsets that are used by the operator to 'teach' each robot a variety of positional movements and actions. These are recorded by the robot and then played back and repeated in order to carry out the production operation.

Accurate to 0.3 mm, once programmed by the operator, the robot intelligently adapts the program to find the best sequence of movements, therefore reducing cycle times and saving valuable seconds.

In operation the robot provides feedback to the operator through the handset. Also, if a robot stops work for some reason, the robots that come after it on the production line recognise this and continue to work until they are unable to do so due to the lack of work being passed from the 'down' robot.

Nissan also uses a vast range of other control technology. Various components are carried to the manufacturing cells on the production line using conveyors, while overhead cradles are used to position the car assemblies.

Another form of materials handling is the use of automated guided vehicles. AGVs are used to transport components from stock to the appropriate manufacturing cell. They do this by following a wire embedded in the factory floor, which is detected by a sensor on the AGV.

AGVs are used to continuously carry wiring harnesses from a docking station to the production line.

The sight of the robots working with such accuracy and repeatability is quite extraordinary and very impressive, perhaps best summarised by Chris Purnell:

'I've been round this place a hundred times and it is always inspirational to watch the robots at work ... I could watch them all day long.'

Chris Purnell, Training Controller at Nissan UK, demonstrates the control handset with Steve Wallis

case study

Newspaper production

Most people would probably admit to taking a local newspaper for granted – it turns up on time, day after day, it is purchased for a relatively small price, it is read and then used for packing boxes or is discarded and hopefully recycled.

What many people would be surprised by is the amount of new technology and effort required to produce a local daily newspaper.

Take, for example, the *Hartlepool Mail*, which, every day, reaches 21,000 homes in the area, where it is read by approximately 60,000 people.

It is published six days a week and the company (which is owned by a larger business – the Edinburgh-based Johnston Press PLC) has a turnover of approximately £132 million.

When you walk into the printing department of the newspaper you are immediately struck by the amount of new technology being utilised. Overhead conveyors position what seems like an endless stream of paper into a highly automated printing press. The press then uses various cylinders applied with different-coloured inks (black, yellow, magenta and cyan) which roll over the paper, producing the full-colour images you see in the final newspaper. Each cylinder can be individually positioned to a fraction of a millimetre to ensure this 'layering' of colour is as exact as possible.

If you look at a colour newspaper, you will be able to see a cross toward the centre of the spread. This cross is the datum used to align the various cylinders – each individual cylinder prints an individual cross. If the cylinders are aligned correctly then you should see only one final cross.

The press can produce an impressive 45,000 newspapers per hour, which are then carried by a series of overhead, roller and slat conveyors to the despatch department to be 'shipped' into the surrounding communities.

Before the printing stage, there is a great deal of new technology used to produce the aluminium plates attached to the press cylinders and which apply the ink to the paper.

The process of producing the newspaper starts in the newsroom where the reporting staff combine local correspondence and photographs with national press stories wired in from news centres.

Using desktop computers, the stories and photographs are placed into templates where they are formatted and edited ready for printing. The software used is a publishing software package called Adobe InDesign.

If you open a newspaper, you will notice that the front and back pages are on the same piece of paper or 'spread', for example page 1 may be joined onto page 28. This is because of the way the computer file is imposed onto the aluminium printing plates – through a process called 'image setting'.

The completed publishing files are combined using 'speed drivers' to produce the different spreads used to make up the newspaper.

Once approved, this file is then transferred onto a special film (as shown below). Using a laser, the machine 'prints' the image and text onto the film, which is then developed, fixed, washed and dried using a series of chemical baths and water.

The negative film is then placed on a piece of equipment called an 'exposure frame machine' which transfers the desired image to the aluminium plate.

The dark parts of the film protect the plate from an ultraviolet light source, while the transparent parts allow

case study

Newspaper production

the light to penetrate through to the special coating on the plate – this coating reacts to the UV light.

When developed, the exposed parts of the plate are darker and are fixed with the newspaper image (as shown below)

When the plate is attached to the printing press cylinders, the ink adheres only to the dark parts of the plate – as the paper is fed through the press the cylinders roll across, producing the final product.

The application of technology

Having read this unit you will now be in little doubt of the impact of technology on our daily lives.

The development of technology is driving forward constantly, allowing the ideas of engineers and manufacturers to be turned into innovative new products. Wherever you are now, simply look up and take a look round – engineering and manufacturing are evident everywhere. Your footwear and clothing, the food you eat and drink, the cars, motorbikes, cycles and lorries on the roads, video games, computers, DVD, video and midi equipment are all developed through the use of new technology.

You have learned how products can be grouped into six sectors:

- mechanical and automotive
- electrical, electronic, computing and telecommunications
- engineering fabrication
- printing, publishing, paper and board
- food and drink, biological and chemical
- textiles and clothing.

All of these sectors use new technology and advanced processing techniques to varying degrees.

You will also now be aware of the three main classifications of new technology.

Information and communications technology

Perhaps you have used some of this type of technology in the past – wordprocessing software to write letters and assignments, spreadsheets to carry out calculations and databases to store information. It is also likely that you have used e-mail to write to friends and the Internet to visit your favourite websites, and to download games and music.

You will now be aware of how industrial companies utilise this technology to aid and assist the engineering and manufacturing of products.

One of key elements of this category of new technology is computer-aided engineering, which includes the key innovations of computer-aided design (CAD) and computer-aided manufacturing (CAM).

Not many industries these days work without some type of CAD – it could be desktop publishing, the graphical design of food packaging, the highly complex technical drawings often used in electrical and electronic work or the component drawings so common in mechanical and automotive engineering.

Equally widespread is the use of CAM to help turn these highly technical drawings into actual products.

New materials and components

This second of the three categories of new technology is arguable the most important.

Humans have been developing new materials for thousands of years, dating right back to the first metals used by early civilisations. Constant progress has been made throughout history, notably by the Egyptians and Romans, and during the Middle Ages and the Industrial Revolution in the 1800s. However, no developments have moved as rapidly as in the last century.

The development of alloy metals, plastics and composites such as glass fibre underpin all of the new products we take for granted today.

The invention of the microchip not only led to the first home computers but was also key to the progress of automation, robotics and control technology used to manufacture many highly complex products. Microchips are now found in many home appliances, including video recorders, DVD players, microwaves and even some toasters and children's toys.

Control technology

The advancement of the final category, control technology, raises many issues regarding the disadvantages of new technology.

With highly automated production lines now used throughout the world to carry out manufacturing processing, questions inevitably arise regarding employment opportunities and the environmental issues associated with the use of new technology.

This is a highly complex issue and varies from one industry to another and from company to company – some of the key concerns are discussed below.

The environmental impact

In the past, manufacturing and engineering activities have contributed to many of the environmental problems that we are seeing today – the depletion of the ozone layer, global warming and acid rain – all due to high levels of water and air pollution.

Some scientists believe that these effects are so devastating that average global temperatures will rise by up to 6°C by the end of the century, heavy rain will cause river deposits to increase by up to 50% and the biggest floating ice sheets on Earth could melt, increasing seawater levels by 7 metres.

So what is being done about it? Industry today is still responsible for the emission of gases such as carbon monoxide and sulfur dioxide into the atmosphere, but the levels are monitored very closely by various environmental groups.

Waste products from older-style industrial plants like this one have had a massive impact on the environment

It is now well understood that waste materials are an inevitable aspect of manufacturing and engineering activities, and it is necessary to dispose of these materials in a safe and responsible manner. Metallic and non-metallic materials, oils, lubricants, scrap products and gases all require safe disposal, treatment or recycling. This is now governed by stringent and detailed international law. Choosing materials that can be safely recycled or disposed of is now a key feature of the design of any new product.

Consider a mobile phone – it contains a large range of materials, including copper, aluminum, ABS plastic and glass.

Key

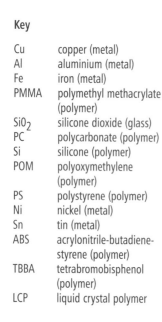

Cu   copper (metal)
Al   aluminium (metal)
Fe   iron (metal)
PMMA   polymethyl methacrylate (polymer)
SiO_2 silicone dioxide (glass)
PC   polycarbonate (polymer)
Si   silicone (polymer)
POM   polyoxymethylene (polymer)
PS   polystyrene (polymer)
Ni   nickel (metal)
Sn   tin (metal)
ABS   acrylonitrile-butadiene-styrene (polymer)
TBBA   tetrabromobisphenol (polymer)
LCP   liquid crystal polymer

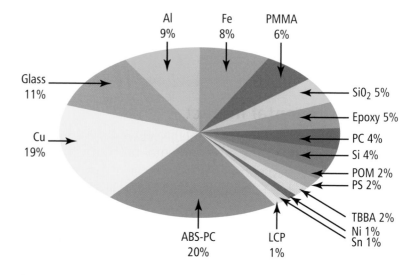

Today, typically between 65 and 80% of the materials used in a mobile phone can be recycled and reused. Furthermore, up to 90% of the plastics materials can be burnt in a furnace and used as fuel.

Manufacturing and engineering have no doubt damaged our planet over the past few hundred years since the evolution of mass production – it is how we proceed from now that will dictate the future of the environment.

The loss of employment opportunities

During the twentieth century, advances in automation have been blamed for the loss of employment in industries that depended heavily on large workforces – the car industry, for example.

Industrial robots do not need to take a break, they can work through the night and never get tired – consequently they have replaced the need for manual labour for many production activities. Machinery such as conveyor belts have reduced the need for manual material handling and CAD has been accused of reducing the level of job opportunities in technical design.

Whilst this is no doubt true, it is also necessary to consider the employment created by new technology. Advances in electronics, microelectronics and new technologies have resulted in new products, which have created vast ranges of job opportunities – ICT consistently proved the largest growing industry sector within the 1980s and 1990s.

It is highly unlikely that manual labour will ever be completely replaced in manufacturing. Many industries such as the textile, food and electronics sectors still place a great deal of importance on people power. As a workforce, people are able to learn quickly, and they can adapt to new skills that can be acquired and refined much faster than the automated equivalent.

Automation can be incredibly expensive to install and its introduction can usually be justified only in mass production, hence there will always be a need for manual labour in engineering and manufacturing.

In conclusion, then, how important is new technology?

You should be in little doubt that it is vital to maintain the standard and quality of living we take for granted in this country today. In order to do this, technology must continue to develop.

Advances are constantly being made in ICT, materials, new components and control technology, and the driving force behind these are the engineers who are designing, developing and manufacturing new products that hit the marketplace every day. If you complete this qualification and move on to choose a career in engineering, you too could become a part of this process.

case study

Mechanical/automotive sector

Car engine

Introduction

One of the key areas of this sector is transport manufacture – in the UK it is worth around £61 billion a year.

An essential part of any vehicle is the engine. The engine's function is to compress and ignite fuel, which produces the necessary drive to enable a vehicle to move.

Key parts

The key parts of an engine include:

- engine block
- pistons
- cam shaft with followers
- crank shaft
- cylinder head
- belts.

Key things to investigate
ICT

- CAE is extensively used in the design and manufacture of all engine components, and this is unquestionably the most important aspect of ICT used in the production of car engines.

- CAD is used to design the parts, at which point the design is transferred to CAM packages to assist in the process design.
- CNC is often used to design casting and moulding tools in addition to performing machining operations to exact specifications.

Materials and components
The principal group of materials used is the metals – to investigate further, look into aluminium alloys and carbon steel alloys.

Polymer materials such as rubber and ABS plastics are also used on many components.

Systems and control technology
With such high product volumes being manufactured in the car industry, many of the casting and forming processes used are automated and computer controlled.

To investigate further, look into modern casting and forming processes and computer-integrated measurement to check quality.

Where to look
Check out the major car manufacturers such as Ford, Peugeot, Nissan, Vauxhall and Volkswagen – all have websites and have documentation available on their production methods.

Engineering fabrication

Mountain bike

Introduction

Used by millions of people worldwide, the mountain bike is now the most popular bicycle in the Western world. Depending on the quality, mountain bike prices start at under £100 and rise to several thousands of pounds for top-of-the-range models.

Popular manufacturers of mountain bikes include Raleigh, Saracen, Orange, Marin and Claude-Butler.

Key parts

The key parts of a mountain bike include:

- frame and forks
- suspension
- handlebars and grips
- gearing
- wheels.

Key things to investigate

ICT

- CAD is often used in the design of mountain bikes to produce technical drawings of components and materials.
- CAM is used to help process the more intricate parts such as the gearing, etc.

- The manufacturers have websites to market, promote and sell their products.

- The manufacturing companies use a database to keep track of their employees, spare parts and bike models.

New materials and components

Mountain bikes consist of different materials and components designed to have specific properties such as good strength-to-weight ratios. Some of the key materials and components for investigation include:

- aluminium alloys and other metals used for the frame
- carbon and stainless steel used for gearing and brake cables
- fixtures such as various screws and bolts
- polymer materials used on the brakes, tyres and handle grips
- composite materials used for state-of-the-art wheels.

Systems and control technology

- In the larger manufacturers, state-of-the-art welding using TIG (tungsten inert gas shielding) and argon methods are automated to high quality standards.
- The painting process also utilises computer control technology to ensure an excellent finish.
- CNC is used to produce press tools and to carry out the final machining of components such as the gear spurs.

Where to look

Check out the websites shown opposite.

Saracen
www.saracen.co.uk

Orange
www.orange.co.uk

Raleigh
www.raleigh.co.uk

Marin
www.marin.co.uk

Shimano
www.shimano-
europe.com/cycling/

case study

Electrical, electronic, telecommunications and computing

Portable CD player

Introduction

Becoming popular in the early 1990s, portable CD players allowed consumers to listen to digital music anytime, anywhere.

They are now extremely compact and lightweight, possess a high quality finish and can be purchased for a fraction of the original introductory price.

Popular manufacturers of these products include Sony, Sharp, Aiwa, Goodmans and Technics.

Key parts

The key parts of a portable CD player include:

- case and cover
- laser lens and pickup
- LCD display
- motor drive mechanism
- PCB
- headphones.

Key things to investigate

ICT

- CAD is used extensively in the design of CD players to produce all the technical drawings of components used on the PCB, drive mechanism and cover.
- CAE is used to help manufacture the parts using CAM and CNC. For example, the PCB is machined using a CNC milling machine and the plastic parts such as the

cover utilise CAE to machine the tools required for injection moulding.

- The manufacturers have websites to market, promote and sell their products.
- Companies use databases to keep track of their employees, spare parts and models.

New materials and components

- Investigate the properties of the glass fibre material used as the PCB base. Also on the PCB, examine the different components used, such as diodes, resistors and integrated circuits.
- Determine the polymer or metal material used as the cover for the CD player – ABS plastic and aluminium are often used.
- Investigate the LCDs – what are they made from, how do they work?
- Research the operation of a laser – how does it read the CD?

Systems and control technology

The manufacture of CD players is highly automated. Some key topics for investigation could include:

- the injection-moulding process used to manufacture the cover and other plastic parts
- surface-mount robotics used to place the tiny components into the exact position on the PCB
- conveyors and transfer machines required to handle parts and move materials between workstations in the factory
- how manual methods are integrated into the process using highly automated assembly procedures.

Where to look

- How Stuff Works (www.howstuffworks.com)
- Sony (www.sony.co.uk)
- Digital Recordings (www.digital-recordings.com)

Google search topics

- surface mount technology
- injection moulding
- PCB assembly
- CD player assembly

case study

Textiles and clothing

Cross training shoes

Introduction

You would be hard pushed to find someone who doesn't own or hasn't in the past purchased a pair of training shoes. Used every day by athletes and sportsmen and women, as well as being a popular item of fashion clothing, the training shoe industry is worth billions of pounds worldwide each year.

A hugely competitive industry, the large manufacturers such as Nike, Adidas, Puma and Reebok pour millions of pounds into advertising each year to remain at the top of the industry.

Key parts

The key parts of a training shoe include:

* outsole (bottom of training shoe)
* midsole (cushioning part)
* upper (top of training shoe).

Key things to investigate
ICT

* Prominent in the manufacture of training shoes is the use of CAE. CAD is used to design, develop and produce 3D models of the training shoe. CAM and CNC are used to manufacture the mould used to produce the outsole and midsole rubber materials. CAE is also used to cut the patterns used for the uppers.

- One of the key ways ICT is used by training shoe companies is for advertising. Take a look at any of the websites in the 'Where to look' section – all use sophisticated graphics and represent state-of-the-art web design. Companies advertise using the Web, by sponsoring sportsmen and women, and on TV. All these advertising methods provide product awareness to the marketplace.

New materials and components
An interesting aspect of the design and development of cross training shoes is the use of new materials.

- Outsoles are typically made of extremely durable carbon rubber or a combination of blown rubber to provide a more flexible, lighter outsole.
- A polymer such as polyurethane is often used as the midsole to make a dense and durable cushioning material.
- The majority of training shoes utilise a leather upper, often combined with a synthetic lightweight mesh to allow the foot to breathe.

Systems and control technology
- The training shoe industry still uses high volumes of manual labour, so it tends to be largely based in Asia where labour costs are very competitive.
- Some of the systems and control technology used includes transfer machines and conveyors, and automatic cutting machines.

Where to look
Check out the following websites:

- Nike (www.nikebiz.com)
- Adidas (www.adidas.co.uk)
- Mizuno (www.mizunoeurope.com)

Magazine

Introduction

Consumers buy millions of magazines each year in the UK. Some are printed weekly, some fortnightly others once a month.

The most popular types of magazines tend to be fashion, sport, culture, television guides or hobby-based reading.

Key things to investigate
ICT

ICT is used widely in the production of magazines. Two of the main examples are:

- Digital photography and text are combined using specialist publishing software – a form of computer-aided design.
- E-mail is used to communicate correspondence and in some cases articles and photographs produced by journalists who may not be based at the magazine headquarters.

Materials and components

Consider the grade and type of paper used. How does it differ from standard A4 in weight and composition?

Systems and control technology

The printing and publishing industry relies heavily on good quality printing presses. These are hugely automated, using different types of conveyors to transport the raw materials and finished product.

Where to look

- The 'British Printing Industries Federation' is a good place to start for information on this industry.
  (www.bpif.org.uk)

Google search topics

- printing processes
- industrial paper supply
- industrial printing

Bottled water

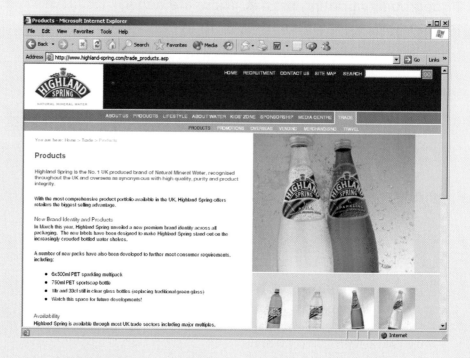

Introduction

You may think 'What's a bottle of water got to do with new technology?' and you would be asking yourself a perfectly reasonable question. However just because water is a natural product it doesn't mean that new technology isn't used at some point in its production.

Key things to investigate
ICT

ICT is used by the manufacturers of bottled water in several ways:

- Correspondence between company members is done via e-mail and the companies also have websites to promote their products.
- CAD is used to design the packaging and labels on the bottles – this is extremely important in developing an image which suggests a top quality product.
- The bottle is designed using CAE and utilises CNC machines to produce moulding tools used to manufacture the finished plastic parts.

Food and drink

case study

New materials and components
The main ingredient is obviously water; naturally sourced from locations such as the French Alps.

Another main ingredient is a recyclable polymer such as polyethylene used in the manufacture of the bottles.

Systems and control technology
Despite producing a natural product, this industry is still heavily automated.

Advanced conveyor systems are used to transport the product to the various stages of bottling, filling and applying the bottle tops. Various types are used, such as roller and slat, which snake round in many different directions to transport the product from stage to stage.

Where to look
- The major manufacturers such as Highland Spring, Evian and Volvic have limited information on their websites – take a look as part of your research. (www.highland-spring.com, www.evian.com and www.volvic.co.uk)
- A good place to find information on the types of conveyors used is the manufacturers – run a search on 'conveyor manufacturers'.
- www.isomaconveyors.co.uk and www.armax-conveyors.co.uk have some very good information.

Exam hints

1 Read all the questions first and determine which carry the most marks and which questions you are confident in answering. With this in mind, attack these first, making sure you plan your time correctly and earn all the marks on offer.

2 Don't forget what is included in the three major categories of ICT, materials and components, and systems and control technology.

3 Read the questions carefully – a common mistake made by students is that a question may, for example, ask about control technology and the student's answer is about CAD – this is an example of ICT, not control technology. If the questions are asking about ICT/new materials/control technology, make sure your answers cover the relevant technologies within these categories – include as much information as you can, as long as it is relevant. To help you avoid misreading a question, underline the key words first, for example:

Describe <u>three advantages</u> of <u>control technology</u> used to make a <u>mobile phone</u>.

This should remind you of exactly what to include in your answer.

4 If the question is referring to a product, make sure you apply your answers to that product – try not to get sidetracked into discussing other examples.

5 Don't rush, and use all of your time to ensure you have answered the question fully and as best you can.

6 Think carefully about answering the question before you put pen to paper – use some scrap paper to jot down a few notes if necessary.

7 Write neatly and carefully; check your spelling and grammar.

Revision questions

1   Draw a line between the products shown below and their appropriate sector of engineering and manufacturing.

    Mechanical/automotive   Laptop computer
    Engineering fabrication   Mountain bike
    Electrical/electronic/computer   Football boots
        and telecommunications   Shower gel
    Textiles and clothing   Audi TT
    Printing and publishing/paper and board   Hardback book
                                                     CD player
                                             Bottle of cola drink

2   State an application of ICT in engineering and manufacturing.

3   State three benefits of CAD to a company.

4   State three disadvantages of CAD to a company.

5   Why is the Internet important to engineering and manufacturing companies?

6   How have robotics improved efficiency in the workplace?

7   What is a PLC and how is it used in industry?

8   State a traditional method of working in manufacturing and the technology that has replaced it.

9   What environmental consequences can result from poor manufacturing methods?

10 Has new technology affected the workforce in engineering and manufacturing?

Glossary

aesthetics: the appearance and form of a product or component

alloy: a mixture of two or more metals or elements

AlphaCAM: a piece of software used to carry out computer-aided manufacture (CAM) functions

analogue: a signal that is continuously variable, e.g. temperature or height

assemble: to put the individual parts of a product together

British Standards: a set of instructions and guidelines used to carry out engineering activities to the correct level of performance and quality

CAD: computer-aided design. It involves the use of computers to carry out design work that used to be carried out manually

carbon steel: a mixture (or alloy) of iron and a small amount of carbon. It is commonly used in engineering

cavity: the recess or imprint made in sand into which molten metal is poured during casting

communication: the verbal or non-verbal exchange of information

computer-aided design (CAD) systems: systems that use computers to design components and therefore allow you to view a component graphically before it is machined

concept drawing: an initial sketch to convey design ideas and solutions

convention: an agreed method of representing a feature

co-ordinates: two or three distinct numbers that provide an exact position in a given area or space. (30, 40) would indicate 30 along in a horizontal direction (sometimes referred to as the x-axis) and 40 high in a vertical direction (the y-axis). A further axis (z-axis) is sometimes used to specify depth

core: the centre of an inductor. It is generally iron, or where there is nothing in the centre of the inductor, air

critical control point: an important point during production when a component should be checked or inspected

design specification: a list of conditions that must be met. It takes into account the original design brief but also takes into account the decisions made about the key features

digital: a signal that can be in only in two states, e.g. a light switch, which can be on or off

ductile: can be stretched and deformed without breaking. This property is known as ductility

ductility: the ability of a material to be pulled and formed into a desired shape without breaking or fracturing

eutectoid: where steel has a carbon content of 0.8%. Steels with less than 0.8% carbon are known as hypo-eutectoid steels. Steels with more than 0.8% carbon are called hyper-eutectoid steels

farad: the unit for measuring capacitance, named after the scientist Michael Faraday

fatigue: a form of material failure that is often caused by repetitive, cyclic stress – it often causes cracking of a component

fax machine: an abbreviation for facsimile machine – facsimile simply means exact copy or reproduction

file card: a flat wire brush used to clean hand files

filler: used to assist in the welding operation by providing extra material to join parts together

filler material: the material that is melted to fill the gap between two pieces of metal that are to be joined together during non-fusion welding

flow chart: a special diagram that can help direct the reader to a product, service or problem

formability: a measure of how well a material can be formed

foundry: a place where metal is melted and poured into moulds to form castings

grains: the name given to the particles that steel is made of

hand tools: the general tools in the workshop that do not need power

hardware: any part of the computer that you can physically touch, e.g. monitor, keyboard, hard drive, etc.

heat treatment: the process of changing the properties of metals

hertz: the standard unit used to measure frequency, named after Heinrich Hertz, a pioneer in this field of science

homogeneous: a material that is made up of only one material, not a mixture

IEE: the Institution of Electrical Engineers is an international organisation for electronics, electrical, manufacturing and IT professionals

infrastructure: the way in which communication is linked and organised

intranet: an internal network of computers, usually within a business, school, college or other medium to large organisation

ionisation: when a gas becomes able to carry electricity, i.e. it becomes a conductor of electricity

key features: the important individual parts of a product that need to be considered

marking out: drawing on metals using special ink and steel marking-out tools

mind mapping: a way of graphically representing ideas and linking them together

model: a computer-generated graphic in 3D. It allows the user to scroll to different views to examine all features

non-fusion welding: when the two pieces of metal to be welded do not melt. Only the filler material melts which holds the metals together

nuts: these can be hexagonal, square or any number of shapes with an internal thread for use with bolts

ohm: the unit for measuring resistance, named after the mathematician Georg Simon Ohm

parent metal: the main pieces of metal that are to be bonded together during welding

peaks and valleys: terms used to describe the highest and lowest points on the surface finish of a component

plant: a term used in engineering to describe a complex system of equipment designed to produce a specific outcome – a power plant, for example

process: the action of doing a task. It could be by an operator or a machine

product specification: a list of all the important features of a product

productivity: the amount produced, or work done, for a given cost

prototype: a model of a product made before production to help designers

quality: a measure of how well a product does the job that it is designed to do

quality manager: a person responsible for the overall quality of the products produced by a business

risk assessment: a method of looking at a particular area or procedure and determining what risks are present. A risk assessment form will be produced as guidance for safe practice

sample: a component or product that has been specially produced or selected to represent the acceptable quality for manufacture

software: the programs that run on the computer. A well-known example would be Microsoft Word

supplier: any business, shop, or person that provides materials to another business or customer

swarf: the waste material that comes from machined material. It is usually very sharp and can be hot

tensile test: stretches a material to see how well it stretches. This stretching is known as elongation

turnover: the total sales achieved by a company or organisation

weld bead: the joint created that holds two materials together during fusion welding

weldability: the ability of materials to be fusion welded

Index